"A timely and much needed book given how far and
changed over the last few years. The ability to adap
we work is now one of the key drivers of organisational and individual performance
and resilience. The death of the office may have been greatly exaggerated, but it
has emerged from the paroxysms of the last few years with a new and constantly
evolving role. Nigel Oseland has set out clearly, with plenty of examples, how
managers can gauge where they are, their direction of travel and the ways in
which they might pause, change course or accelerate to deliver a better workplace
experience for people and better results for the organisation."
  – **Mark Eltringham**, *Publisher of IN Magazine and workplaceinsight.net*

"If you have more than 20 people on your team, either a corporate, academic, or
other type of organisation, I highly recommend this book. It's our go-to practical
guide for conducting POE research, arguably the single most important task in
workplace change projects. With in-depth advice and invaluable case studies it's an
essential resource for designing high-performance workspaces."
  – **Alenka Kragelj Erzen**, *CEO Kragelj Architects*

"Oseland uncloaks the complexities of POE and, again, has produced an
informative, easy-to-read, go-to resource. He emphasises POE as a critical aspect
of the evidence-based design approach and reminds us to move beyond the view of
'what we got wrong' and, instead, embrace how we can do better. In the changing
world of the workplace, from hybrid working, health and wellbeing and Carbon
Net Zero, it is clear the time for permanent and consistent use of the POE has
come."
  – **Paige Hodsman**, *Concept Developer, Workplace Specialist,*
Saint-Gobain Ecophon

"Following the publication of this book, there can no longer be any excuse for not
understanding the role or value of POE. On the one hand, this book is comprehensive
and hugely informative, while on the other hand it is highly readable to both
technical and lay readers. The book is necessarily highly structured, and delves
into the detail of POE methodologies, but Nigel's writing style is welcoming and
largely non-technical."
  – **Rob Harris**, *Principal, Ramidus Consulting*

"This book spells out so many benefits of POE, with examples of varied approaches
to meet the range of client objectives. With the cost of POE so marginal, the
proposition for it is persuasive – a stepping stone to enhanced effectiveness."
  – **Ziona Strelitz**, *Founder Director, ZZA Responsive User Environments*

"This book discusses post-occupancy evaluation through a wide range of viewpoints. It covers the multifaceted nature of office buildings and the methodology available for evaluation processes from both user experience and building performance points of view. Nigel Oseland's practical and experienced approach to the subject provides valuable guidelines for setting up and fine-tuning post-occupancy evaluation processes."

— **Piia Markkanen**, *Doctoral Researcher, Oulu School of Architecture, University of Oulu*

"Finally, we have a handbook of practical guidance on conducting post occupancy evaluation. Dr Oseland dusts off what can be a dry subject and makes it sexy again. He dispels the myths surrounding POE and demonstrates the real value it can bring to an organisation. Particularly interesting is the recommendation of an annual POE.

— **Maggie Procopi**, *Founder, Workplace Trends*

A Practical Guide to Post-Occupancy Evaluation and Researching Building User Experience

A Practical Guide to Post-Occupancy Evaluation offers high-level pragmatic guidance and case study examples on how to conduct a Post-Occupancy Evaluation (POE) to determine whether a workplace project is successful and uncover the lessons learned for future projects. For designers, POEs provide essential pre-design feedback, informing the design brief to determine occupant requirements and help focus expenditure. For those in charge of a building or buildings, POE offers proactive building management and can also be used as part of the change management programme in larger projects, informing the occupants of progress. The practical guidance offered in this book will help the workplace industry understand if a design meets the requirements of an occupier and measure the success of and value offered by a workplace project.

This book will be of interest to professionals in the workplace industry responsible for delivering and evaluating capital projects as well as those studying interior design, architecture, surveying, facilities management and building services engineering.

A Practical Guide to Post-Occupancy Evaluation

and Researching Building User Experience

Nigel Oseland

Routledge
Taylor & Francis Group

LONDON AND NEW YORK

Designed cover image: © Getty Images

First published 2024
by Routledge
4 Park Square, Milton Park, Abingdon, Oxon OX14 4RN

and by Routledge
605 Third Avenue, New York, NY 10158

Routledge is an imprint of the Taylor & Francis Group, an informa business

British Library Cataloguing-in-Publication Data
A catalogue record for this book is available from the British Library

Library of Congress Cataloging-in-Publication Data
Names: Oseland, Nigel, author.
Title: A practical guide to post-occupancy evaluation and researching
    building user experience/Nigel Oseland.
Description: Abingdon, Oxon; New York, NY: Routledge, 2024. | Includes
    bibliographical references and index.
Identifiers: LCCN 2023012492 (print) | LCCN 2023012493 (ebook) |
    ISBN 9781032396651 (hardback) | ISBN 9781032390925 (paperback) |
    ISBN 9781003350798 (ebook)
Subjects: LCSH: Office buildings — Evaluation. | Offices — Evaluation. |
    Work environment — Evaluation. | Facility management.
Classification: LCC HD1393.55. O84 2024 (print) | LCC HD1393.55
    (ebook) | DDC 658.2/3–dc23/eng/20230323
LC record available at https://lccn.loc.gov/2023012492
LC ebook record available at https://lccn.loc.gov/2023012493

ISBN: 978-1-032-39665-1 (hbk)
ISBN: 978-1-032-39092-5 (pbk)
ISBN: 978-1-003-35079-8 (ebk)

DOI: 10.1201/9781003350798

Typeset in Times New Roman
by Apex CoVantage, LLC

Contents

Author biography

Dr Nigel Oseland is an environmental psychologist, workplace strategist, change manager, researcher, international speaker, lecturer and published author with 11 years of research and 25 years of consulting experience. In his consulting business, he draws on his psychology background and his own research to advise occupiers on how to redefine their workstyles and rethink their workplace to create working environments that enhance individual and organisational performance and deliver maximum value. Nigel's design projects are evidence-based, and he conducts a post-occupancy evaluation (POE) as part of each project.

Since 1996 Nigel has conducted over 100 POEs using his own and other organisations' methodologies. He authored the British Council for Offices *Guide to Post-Occupancy Evaluation* in 2007. Nigel facilitates regular training workshops on POE, attracting participants from around Europe, some employed in the corporate world and others working in the public sector or for higher education institutions. He also lectures on POE at The Bartlett, University College London, and has presented on POE at conferences around the globe. Nigel also created the original Wikipedia entry on POE.

Nigel has published over 150 academic articles and has contributed to chapters in numerous books, including *Architecture Beyond Criticism: Expert Judgment and Performance Evaluation* (Preiser et al., 2015) and *Building Performance Evaluation: From Delivery Process to Life Cycle Phases* (Preiser, Hardy and Schramm, 2018). Nigel's most recent book, *Beyond the Workplace Zoo: Humanising the Office* (Oseland, 2022), emphasises the importance of POE.

Nigel spends his spare time regularly gravel biking in the nearby Chilterns and occasionally downhill mountain biking in Wales and the Alps. He is a member of his local Toastmasters International speakers club, where he hones his presentation skills. Nigel is a budding beer sommelier and organises the annual local beer festival and other beer-related events. Nigel also presents a weekly show on his local community radio station.

Foreword

Investors and developers spend vast amounts of money delivering new workplaces, supported by an increasingly labyrinthine network of professionals – architects, engineers, surveyors and many, many others – and then lease them to occupiers who, in turn, spend large amounts of money fitting them out, managing them and putting up with their shortcomings. And so it goes on. There is very little feedback from occupier to professional to provider; everybody simply moves on to the next "project".

This is, of course, an oversimplification, but the essence is fair: in real estate, there is virtually no feedback from customer to provider. And yet we hear over and over that occupier satisfaction with buildings is generally poor, that buildings can inhibit productivity, that they can be inefficient and even that they can negatively impact health.

At the same time, the techniques for evaluating building performance have been around for decades. As Nigel notes, his first book on the topic dates from 25 years ago. So there remains a big question around why post-occupancy evaluation (POE) is not as well established as much later creations such as BREEAM and LEED, which set standards in design and specification. But following the publication of this book, there can no longer be any excuse for not understanding the role or value of POE.

On the one hand, this book is comprehensive and hugely informative, while on the other hand, it is highly readable to both technical and lay readers. The book is necessarily highly structured and delves into the detail of POE methodologies, but Nigel's writing style is welcoming and largely non-technical. The book breaks down into three main areas: rationale (what is a POE, why undertake one and when is one appropriate); method (who and how to survey); and output (reporting).

Importantly, and as you might expect from a qualified psychologist, Nigel's book does more than just address the relationship between worker and physical environment: it also looks at the importance of organisational performance and cognitive and physiological monitoring. The book also explores how to include sustainability audits within POEs. In fact, a key message from the book is that there is not one preferred way of undertaking POEs, but instead, and depending upon purpose, audience and scope, there are numerous ways of approaching POEs.

In the latter part of the book, Nigel provides some very useful advice on how to present what is often fairly dry data in informative and easily understandable ways.

Even before the Covid-19 pandemic, the use of office buildings was evolving rapidly away from factory-style environments to much looser fit-outs and agile workstyles, reflecting the use of digital technologies and new work processes. The pandemic has turbo-charged this process such that "hybrid working" – part in the office, part at home, part elsewhere – has become normal. As a consequence, workplaces (and employers) will have to work harder to attract staff to the office environment. In these circumstances, it will be doubly important to gather and act upon feedback from occupiers.

Rob Harris
Ramidus Consulting
January 2023

Acknowledgements

A book like this one is not written without support and encouragement. I especially thank Maggie Procopi for that and for the review and continuous flow of cups of tea. Thank you again to Melanie Thompson for copyediting the draft manuscript. I am indebted to Paul Bartlett, my old colleague and dear friend who sadly passed away, for introducing me to performance evaluation. Special thanks to Richard Kauntze, Rob Harris and the British Council for Offices team for permitting me to rewrite the original *Guide to Post-Occupancy Evaluation*. Finally, much appreciation goes to all those who submitted a case study: Aja Ceglar, Bostjan Erzen, David Lehrer, Gary Raw, Mark Eltringham, Piia Markkanen, Susanne Colenberg, Tim Oldman and Ziona Strelitz.

Acronyms and abbreviations

AI	artificial intelligence
AIA	American Institute of Architects
ASHRAE	American Society of Heating, Refrigerating and Air-Conditioning Engineers
BCO	British Council for Offices
BIFM	British Institute of Facilities Management
BIM	building information modelling
BMS	building management system
BOSSA	Building Occupants Survey System Australia
BPE	building performance evaluation
BRE	Building Research Establishment
BREEAM	Building Research Establishment Environmental Assessment Method
BS	British Standard
BSI	British Standards Institution
BSRIA	Building Services Research and Information Association
BUS	Building Use Studies
CABE	Commission for Architecture and the Built Environment
CBE	Center for the Built Environment, University of California, Berkeley
CIBSE	Chartered Institution of Building Services Engineers
CO$_2$	carbon dioxide
DQI	Design Quality Indicator
DQM	Design Quality Method
DTI	Department of Trade and Industry
ECG	electrocardiogram
ESM	experience sampling method
FTE	full-time equivalent
GIA	gross internal area
GIS	geographic information system
GPS	global positioning system
GSL	Government Soft Landings
GSR	galvanic skin response
H&S	health and safety

HEFCE	Higher Education Funding Council for England
IAQ	indoor air quality
IEQ	indoor environmental quality
IFMA	International Facility Management Association
IPA	Infrastructure and Project Authority
IPD	Investment Property Databank
ISO	International Organization for Standardization
IWBI	International WELL Building Institute
IWFM	Institute of Workplace and Facilities Management
KPI	key performance indicator
kWh	kilowatt-hour
LEED	Leadership in Energy and Environmental Design
Lmi	Leesman Index
NABERS	National Australian Built Environment Rating System
N	number
NIA	net internal area
NLA	net lettable area
NOA	net occupiable area
NSW	New South Wales
NUA	net usable area
OGC	Office of Government Commerce
OLS	overall liking score
OPN	Office Productivity Network
PAS	publicly available specification
PMV	predicted mean vote
POE	post-occupancy evaluation
PROBE	Post-occupancy Review Of Buildings and their Engineering
QR	quick response
RAT	Remote Associates Test
RIBA	Royal Institute of British Architects
RICS	Royal Institution of Chartered Surveyors
ROWI	return on workplace investment
RR	response rate
SMART	specific, measurable, achievable, relevant and time-bound
temp. unocc.	temporarily unoccupied
TOCS	Total Office Cost Survey
TVOC	total volatile organic compounds
UK	United Kingdom of Great Britain and Northern Ireland
US	United States of America
UX	user experience
VOC	volatile organic compound
WELL	WELL Building Standard

Epilogue

Whenever you buy a product or service online, purchase a new car or take a taxi ride, pass through security at the airport, stay at a hotel, grab a coffee at your local café or go to a public toilet, the chances are that you will be asked to give feedback on your experience. Seeking feedback is particularly commonplace in the retail and leisure sectors. Customer satisfaction is regularly evaluated, and the information is used to identify and correct immediate faults, improve products and services in the short term, and plan ahead for new products and services that customers will desire in the future. Ultimately, this proffers commercial advantage, creates more sales and increases profitability.

The same commitment to customers is less evident in architecture, property and the construction industry. Back in 1998, Sir John Egan, in his prominent report *Rethinking Construction* (also known as the Egan Report), noted that "the construction industry tends not to think about the customer . . . Companies do little systematic research on what the end-user actually wants, nor do they seek to raise customers' aspirations and educate them to become more discerning. The industry has no objective process for auditing client satisfaction". Considering the pivotal role that office premises play in any organisation, it is unfortunate that the majority of occupiers, designers and facilities managers do not regularly seek feedback from building occupants and ask them what they think the impact of that building might be having on their comfort, wellbeing and performance. As Zimmerman and Martin (2001) memorably pointed out, "Without a feedback loop every building is, to some extent, a prototype – spaces and systems put together in new ways, with potentially unpredictable outcomes".

One well-established system for collecting feedback on the impact of buildings on people is post-occupancy evaluation (POE), introduced by social scientists and building-use experts some 60 years ago. I conducted my first POE around 25 years ago, and I wrote my first guide to conducting POEs over 15 years ago, which was in direct response to the Egan Report (see Figure 0.1). After a decade as a government researcher, I developed a standardised feedback survey, with a paper-based version published for all in the back of *Improving Office Productivity* (Oseland and Bartlett, 1999), and my first consultancy role was to carry out POEs on behalf of a facilities management company. I continue to conduct POEs today and now have over 100, in various forms, under my belt. I have also carried

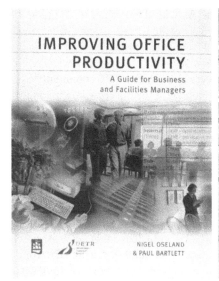

Figure 0.1 My two previous books on evaluating building performance

out numerous one-day workshops instructing the participants on how to perform a POE. So, hopefully, it is evident that I am an experienced advocate of POE. Not only do I use the data obtained to evaluate a building, but I also use POE as part of an evidence-based design process.

The rationale for this book is to revisit the British Council for Offices (BCO) *Guide to Post-Occupancy Evaluation* (Oseland, 2007) that I wrote all those years ago. That guide attempted to "raise awareness of the benefits of POE and to make it accessible to the occupiers, developers and designers by providing practical advice on how they can instigate their own POE studies", and it was made available to all, although primary distribution was to BCO members whereas it would have benefitted from wider circulation.

Today the motivation for POE has changed, and there are more recent innovations in POE methodologies. More importantly, the world of work and workplaces continues to evolve. For example, people are now in the habit of working from home or other spaces, and this is reflected in the lower utilisation of offices. When it comes to satisfaction with offices, occupants are literally voting with their feet. Under-occupied office space, with rows of empty desks, is not sustainable or economically viable. As such, that original 2007 BCO guide is in desperate need of updating to reflect the recent and current situation – so consider this book a revised and revamped version of it.

There are many papers on POE, including case studies, and quite a few textbooks. However, the majority of these publications are lengthy academic and theoretical tomes with assumed knowledge of POE. As with my previous book, *Beyond the Workplace Zoo*, my intention here is to demystify the subject of POE and present it in an easily digestible format. As with the original BCO guide, the focus is on providing clear, pragmatic advice distilled from a range of academic and practitioner resources. Nonetheless, while evidence-based, I should clarify that the guidance on POE presented in this book is ultimately my personal view, which may differ slightly from the views of my peers in the wider POE community.

This book explains how to develop and conduct POE surveys, how to analyse and interpret feedback, and how to share results so that any lessons learned can be put into practice. The book acknowledges that POE incorporates a wide range of techniques and methodologies, including indoor environment measurements, space and occupancy analysis, and energy audits, to evaluate the performance of buildings. However, as a psychologist, I consider the most fundamental element of any POE is using occupant feedback to measure how buildings meet the occupants' requirements; hence the bulk of this book is on using feedback.

There are many POE techniques used in more detailed research studies, so the book sometimes strays into research methods. I acknowledge that not all the techniques presented here will be used in a standard POE, but they are included for the sake of completeness, and, with advances in technology, they are becoming easier to use as part of a POE.

Some think of POE as a complex and specialised process, so they consider it more effective to use established methodologies. Nevertheless, there are some basic techniques that the novice, with some guidance, can use to conduct a robust and useful POE. For example, it is feasible for organisations to design and implement their own in-house feedback surveys, though it is recommended that tried and tested surveys are first considered and learned from.

This book also aims to be valuable for readers who prefer to commission POEs rather than go to the trouble of developing and conducting one themselves. A good grasp of POE methodologies will make for a more informed and cost-effective third-party POE. This book, therefore, includes a compendium of existing methods. It also offers a series of case studies that highlight the practical benefits of POE and how feedback has been used as part of the design process or ongoing performance measurement to provide successful buildings around the world. The case studies also highlight the range of techniques and approaches to POE.

The many benefits of POE are discussed in the first chapter, but perhaps the most significant benefit is the opportunity to learn from one project to the next and from the experiences of others. This can only realistically happen where a systematic approach to gathering feedback is adopted with very similar methods used across all POEs. Ideally, there would be a freely available database of POE results that could be used to inform a wide range of building projects. Some POE methodologies, particularly regarding occupant feedback, have generated large databases that may be used for comparisons, benchmarking and meta-analysis (such as the Leesman Index, the BUS Survey and CBE Occupant Survey discussed in Chapter 11). However, even a unified approach to feedback is unlikely because it would require different organisations to use the same questions and response scales, and to submit their confidential and possibly sensitive data to a central database. Instead, I have reviewed numerous questionnaires that are regularly used to evaluate the performance of buildings and compiled a list of common themes and questions. Other POE techniques, such as environmental measurements, tend to adopt metrics and measurement techniques that are detailed in standards. However, different standards are adopted, which makes comparing such metrics more difficult than might be expected.

Despite POE having been practised, in one form or another, for around sixty years, and despite the benefits of POE, the take-up has been slow. Nevertheless, for reasons discussed later, POE has begun to make an impact on the agenda of occupiers, designers and developers, not only in the UK and the US but throughout the world. Hopefully, the practical guidance in this book will encourage more organisations and individuals to adopt POEs, and using the sets of common questions highlighted later in this book should make the comparison of buildings easier. Note that sharing information is not just about praising the best-performing buildings. It is just as important to share lessons learned about those aspects of buildings that are performing poorly, although admittedly, such sensitive information can be difficult to share. The key point is to gather feedback on the building's performance and share it externally so that building design and operation can improve, ultimately providing better buildings for organisations in the future.

Anyone interested in improving the quality of buildings will benefit from conducting or commissioning POEs. This includes building occupiers who want better value and to get more out of their building(s) and people. It applies to all construction industry professionals involved in fit-out and refurbishment, especially architects, interior designers, workplace strategists, facilities managers, those in corporate real estate, quantity surveyors, building services engineers and acousticians. Some developers are also expressing an interest in how POE may be used to inform the design of their future developments, and agents and landlords also want to prove how their buildings offer better quality and higher-value facilities.

In the following chapters, I will explain POE using the classic "six Ws" interrogative words: why, what, when, who, where and how. Why a POE should be carried out is the most important W for me, but I need to first explain what POE actually is.

Figure 0.2 The six Ws

1   What is POE?

I appreciate that many readers of this book most likely have a sound understanding of what POE is, but I will explain it nonetheless because there is confusion among the long-term adopters of POE as well as those new to the field.

I am not a fan of the rather ambiguous phrase "post-occupancy evaluation", but it is one that has been used for some time, and I do not want to confuse matters further with new phrases. So, to help prevent any further misunderstanding, this first chapter clarifies what I believe POE actually means and also provides a brief history of how POE came about and became recognised, by some, as an important part of the building design process.

Origins of POE

POE has been around in one form or another for at least 60 years, but it is less clear when the term "post-occupancy evaluation" was formalised and first introduced. One candidate for the origin of POE is Tom Markus and his team at the Building Performance Research Unit (BPRU), based at the University of Strathclyde. They appraised fifty or so schools in Scotland in the late 1960s, and Markus wrote his seminal paper "The role of building performance measurement and appraisal in design method" for the *Architects Journal* in 1967 followed by his book *Building Performance* in 1972. At a similar time, the Royal Institute of British Architects (RIBA) was developing its process for design and construction, originally launched in 1963 as a fold-out sheet but then published as the *RIBA Plan of Work* in 1964. The original plan included "Stage M Feedback", a basic form of POE in which architects and their clients inspected and discussed any problems with the building. A good start, but regrettably, stage M was dropped from later versions of the plan. However, the 2013 restructured *RIBA Plan of Work*, discussed in Chapter 2, does include POE in "Stage 7 Use".

In the US, POE started with one-off case studies in the 1960s and then progressed to system-wide evaluations of university dormitories, housing and hospitals. Van der Ryn and Silverstein's (1967) environmental analysis of Berkeley dorms is an early example of POE, as their comment at the time illustrates: "There is no feed-back channel between planning assumptions and building use". McLaughlin's article "Post-occupancy evaluation of hospitals", published in the *American Institute*

DOI: 10.1201/9781003350798-1

of Architects Journal (1975), is a contender for the first use of the term "POE". Apparently, the following year the AIA Research Corporation commissioned Connell and Ostrander (1976) to write a review of POE methodologies. Following that, the US Department of Housing and Urban Development (1977) published a bibliography of POEs in the report *Post Occupancy Evaluations of Residential Environments*.

During the 1980s, POE fell off the business agenda and was dropped from the architectural curriculum. Meanwhile, it was taken up and developed by social scientists, who made it a more systematic and rigorous process. I will clarify later how there is a recent resurgence in POE, thanks to the introduction of Soft Landings, building information modelling (BIM) and building certificates such as BREEAM, LEED and WELL.

Definition(s) of POE

Wikipedia states that POE is:

> the process of evaluating buildings in a systematic and rigorous manner after they have been built and occupied for some time.

I am quite familiar with the Wikipedia entry because I was the first to post it many years ago. The definition is robust and has stood the test of time: others have added to the Wikipedia page, and the definition is often quoted by authors of POE studies. The definition is a quotation from Preiser, Rabinowitz and White's 1988 book *Post-Occupancy Evaluation*, seemingly the first full textbook on the subject.

The definition by Preiser et al. implies two important aspects of POE. Firstly, POE is methodical and evidence-based rather than simply based on hearsay, anecdotal evidence, intuition or assumption. The definition implies data should be collated from a representative range of reliable sources. Secondly, the building project must be complete and occupied. In other words, it is not the critique of the building at the design or planning stage nor is it an assessment of an empty/unoccupied building. The latter, in effect a pre-occupancy survey, is often undertaken at the building selection stage of a project. For example, an organisation may be considering moving to new premises, so examines several buildings to assess their suitability based on size, location, lease, local amenities, transport nodes and architectural features such as daylight ingress, access, building services, contiguity of space and sub-divisibility. Preiser, a pivotal advocate of POE, and his colleagues often focused on the evaluation of new buildings, but of course, POE can also apply to refurbishment and is usually concerned with the fit-out of spaces rather than just the base-build.

Friedman, Zimring and Zube (1978) highlighted the purpose of POE as "an appraisal of the degree to which a designed setting satisfies and supports explicit and implicitly human needs and values of those for whom a building is designed". Similarly, Preiser (1994) commented that "A building's performance indicates how well it works to satisfy the client organization's goals and objectives, as well as the needs of individuals in that organization". Later, Preiser and colleague Jaqueline

Vischer (2005) expanded on this, stating that POE "addresses the needs, activities, and goals of the people and organizations using a facility, including maintenance, building operations, and design-related questions. Measures used in POEs include indices related to organizational and occupant performance, worker satisfaction and productivity, as well as the measures of building performance". These statements imply that POE will require feedback from the occupants to determine whether the building is performing to their satisfaction. Preiser also distinguished between the requirements of the organisation (employer) and those of the individual (employee).

Initially, RIBA defined POE as "a systematic study of buildings in use to provide architects with information about the performance of their designs and building owners and users with guidelines to achieve the best out of what they already have" (RIBA, 1991). More recently, in the *RIBA Plan of Work 2020* (RIBA, 2020a), POE is considered an "evaluation undertaken once the building is occupied to determine whether the project outcomes and sustainability outcomes set out in the project brief, or later design targets for building systems, have been achieved". RIBA, therefore, extended its definition to include whether or not environmental performance targets are met in practice, but this was linked more to mechanical systems testing than to occupant feedback. The RIBA definition represents a more recent and growing view of the importance of POE in helping to minimise the "performance gap", the difference between a building's actual performance and that expected.

In summary, POE is clearly a systematic process for measuring the building's performance that:

- is conducted once the building has been in operation and occupied for some time, rather than at the building selection, design, construction or commissioning stages,
- is methodical and thorough, based on representative and quantified data, rather than anecdotal evidence or the opinions of a select few or the designer,
- provides feedback on how successful the building is in supporting the occupying organisation, addressing individual requirements and meeting the design brief,
- fundamentally involves feedback from the occupants but may require accompanying more technical evaluations,
- can also be used to provide a comparison of how the building performs compared with predictions.

The original definitions of POE refer to the assessment of "buildings". POE can be used to measure the performance of all types of buildings, including healthcare, residential and educational settings, but is more often used to evaluate workplaces, particularly offices. As more organisations adopt agile or hybrid working, a comprehensive POE of the workplace will also need to capture the views on working from home or elsewhere outside of the office. POE may be used to measure the impact of all types of office projects, including:

- relocations and a move to a new office,
- refurbishment or renovation of an existing office,

- small internal moves and space reconfiguration (churn),
- introducing new technologies and new building systems,
- implementing new facilities management initiatives and organisation-wide programmes,
- changing work processes and office use, such as agile working.

POE terminology

Does post-occupancy mean post-project?

The term post-occupancy evaluation is vague, but the "post-occupancy" part is the most confusing. Bordass and Leaman (2005) interviewed members of the Confederation of Construction Clients and found that "the name 'POE' did the activity no favours: POE was seen as too academic, and too late to benefit the project concerned". Many years ago, Denise Jaunzens, a colleague of mine at the Building Research Establishment (BRE), asked facilities managers what they thought POE meant. Their responses clearly indicated some misunderstanding because some considered post-occupancy to mean:

- after the occupants have left and vacated the building leaving it empty,
- like a post-mortem, such as a study of why a building no longer functions,
- only a post-project, post-move or post-completion survey and not a pre-project or ongoing evaluation.

Interpreting "post-occupancy" as something that happens after the occupants have left is contrary to the objectives of POE, which predominantly relies on finding out the impact of the building on the *current* occupants (see Figure 1.1). Assessing a recently unoccupied building might be carried out as part of the dilapidations valuation, but otherwise, it is of little use.

Not just post-project *Not* once empty *Not* post-mortem

Figure 1.1 What POE is not

Source: Aukett Swanke

The second response may have its grounding in sick building syndrome and is not as ridiculous as it first sounds. However, a POE is more akin to a health check than a post-mortem. The aim is to use the evaluation process to revive and improve the building. To reiterate, POE is simply the evaluation of an occupied building, so perhaps the term "occupied building evaluation" or "in-use evaluation" is more appropriate.

The final response is a misunderstanding worthy of further exploration. If a POE is considered an evaluation of any occupied building, and not just a newly occupied build-ing after a move, then it might actually be conducted prior to a project, where "project" may refer to a move or refurbishment or some other workplace intervention. Such a pre-project/pre-move survey can assess the existing building conditions and performance, how the building currently meets occupant requirements and how it currently performs compared with what was expected. It may also be used to determine a baseline measure prior to a project, which allows for a better pre- and post-project comparison. Comparing the results of a pre- and post-project evaluation is a much more robust approach to assess-ing the success of a project than conducting one survey only after the project has been completed, or attempting to benchmark the results with other, possibly external, surveys.

Alternatively, a pre-project survey may be used as part of the design briefing pro-cess: for example, to determine what the occupants like and dislike about their current building that can be resolved in their next one or to help prioritise the occupants' needs for future buildings. A pre-project review may also be used as part of the business case: for example, the Leesman Index team noted, "many of our clients use the survey find-ings to establish the business case for a major strategic project" (Leesman, 2015).

It is, therefore, appropriate to distinguish between "pre-project evaluation" and "post-project evaluation", both carried out in occupied buildings. The terms "pre-project evaluation" and "post-project evaluation" seem much clearer when refer-ring to a building related project.

Some authors, such as Heath et al. (2019), decided to use "pre-occupancy evalu-ation" to mean "the first part of the POE, to gather the data before any changes have taken place", adding to the existing confusion. However, they did at least acknowledge the significance of a pre-project evaluation. Sailer et al. (2010) also use the term "Pre- and Post-Occupancy Evaluation (PPOE)", adding further ambi-guity. The terms "pre-project evaluation" and "post-project evaluation" seem much clearer when referring to a building related project.

Rather than a one-off pre-project or post-project survey, the evaluation of an occu-pied building may be ongoing. For example, it may be carried out periodically, perhaps annually, like a car MOT test, to determine whether the occupants of a long-standing building still find that it meets their requirements. Such a regular review is particularly useful to facilities managers responsible for operating and maintaining buildings and keeping their customers satisfied. If there is a limited budget, pre-project and ongo-ing surveys help to determine any specific maintenance requirements or other building design elements most in need of resolving from the occupants' point of view.

Building performance evaluation

There is more guidance on "When" to conduct a POE in Chapter 3, but regarding ongo-ing evaluations, I need to introduce and explain building performance evaluation (BPE).

Many researchers now prefer the term BPE and have suggested that POE is part of it. For example, Preiser and Vischer (2005) summarised BPE as "a way of systematically ensuring that feedback is applied throughout the process so that building quality is protected during planning and construction and, later, during occupation and operations". Likewise, the Building Services Research and Information Association (BSRIA) defines BPE as "a form of POE which can be used at any point in a building's life to assess energy performance and occupant comfort and to make comparisons with design targets" (cited by Fletcher and Satchwell, 2015) for new, existing and refurbished buildings.

The BPE concept was fully explained by Preiser and Schramm (1997), who considered it to have six phases across the building life cycle (see Figure 1.2), starting with strategic planning, going through to occupancy and ending with adaptive reuse. Each phase is accompanied by a feedback loop which informs the next phase of the building life cycle, starting with an effectiveness review and ending with a market/needs analysis. "Phase 5: occupancy" and "Loop 5: post-occupancy evaluation" are clearly the ones most relevant to this book.

Figure 1.2 The process of building performance evaluation (Preiser and Schramm, 2005)

Regarding occupancy, Preiser and Schramm (2005) noted that "the BPE approach, with its reliance on feedback and evaluation, maintains a long-term perspective by including the period of occupancy in order to improve the quality of decisions made during the earlier phases" and "During this phase, there is fine-tuning of the environment by adjusting the building and its systems to achieve optimal functioning for occupants". They stated that the POE loop is to "provide feedback from users on what works in the facility and what needs improvement. POEs also test some of the hypotheses behind key decisions made in the programming and design phases. Alternatively, POE results can be used to identify issues and problems in the performance of occupied buildings and identify ways to solve these".

Preiser and Schramm's (2005) BPE model also includes Loop 4: commissioning, which comes after the construction phase and is labelled "post-construction evaluation" in earlier versions of their model. They proposed that "at the end of the construction phase, inspections take place . . . As a formal and systematic review process, this loop is intended to insure [*sic*] that owners' expectations, as well as obligatory standards and norms, are met in the constructed building". This statement emphasises that POE is separate process to a commissioning stage assessment.

Preiser and Vischer (2005) also proposed that "BPE is the process of systematically comparing actual performance of buildings, places and systems to explicitly documented criteria for their expected performance". The recent focus of RIBA and building services professionals on conducting a POE to test that predicted energy and other sustainability targets have been realised in practice is in line with Preiser and Vischer's proposal.

Although the above examples suggest a clear distinction between POE and BPE, there are overlaps, and for some, the division is not so clear. As RIBA (2016) noted, "The definitions of POE and BPE vary slightly across the construction industry and the RIBA does not intend to add to this confusion by making its own definitions but to rather provides guidance on POE/BPE activities as a whole".

What's in a POE

Before choosing the components of a POE and the associated approach, it is crucial to decide what information is required and, more importantly, how it will be used. Bear in mind that, fundamentally, a POE should determine whether the building can "satisfy the client organization's goals and objectives, as well as the needs of individuals in that organization" (Preiser, 1994).

Key performance indicators and other measures of success

Many organisations use key performance indicators (KPIs) to measure and monitor their success over time or to benchmark and compare themselves with their competitors. KPIs often directly reflect the broader vision and objectives of the organisation or, at least, the qualities the organisation considers most important. KPIs generally apply to all branches of an organisation, including production, finance, HR, marketing and sales and may be used to measure its activities such as projects, programmes, products and other initiatives.

The use of KPIs is well established in the construction industry, and they are used to evaluate and compare the progress of individual companies across the sector as well as industry-wide progress. For example, the UK body Constructing Excellence (2018) encourages its members to measure and benchmark metrics under the headings: economic indicators, client satisfaction, contractor satisfaction, profitability, predictability, respect for people, and environmental indicators. The KPI categories include objective metrics such as project costs, time to completion, staff turnover, absenteeism, accident incident rate, energy and water use, and waste removed. Similarly, the UK government adopted the following KPI grouping for construction projects: time, cost, quality, client satisfaction, client changes, business performance, health and safety (DETR, 2000).

The KPIs used by the construction sector underpin the metrics and methodologies that may be adopted by occupiers, developers and designers when evaluating a completed and occupied building (that is a POE). For organisations that already measure success using KPIs, it is advisable to think ahead so that the data collected as part of a POE can also be reported in a form that is compatible with KPIs. It also makes good sense to measure and compare the KPIs pre- and post-project.

The workplace consulting practice DEGW introduced the "three Es" framework as a means of evaluating a building's success (Allen et al., 2004), later endorsed by the Commission for Architecture and the Built Environment and the British Council for Offices (CABE and BCO, 2006). The metrics used in a POE might cover the "three Es" below with an obvious additional fourth E (see Figure 1.3).

1. **Efficiency** – Relates to making economic use of real estate and driving down occupancy costs, so metrics might include cost and space metrics.
2. **Effectiveness** – Refers to how the space is used to support the way that people work, improving output and quality, so measures might include occupant

Figure 1.3 The four Es of building evaluation

feedback, self-reported performance and more objective performance metrics such as staff attrition and absenteeism.

3. **Expression** – Concerns how the building communicates messages to both the occupants of the building and to those who visit it, including branding and the building's iconic status, which may be assessed through customer feedback or an architectural award.

4. **Environment** – Responsible occupiers, particularly those in the public sector, may prefer to include a specific set of sustainability indicators, including energy consumption, water usage and waste.

Neil Usher (2018) proposed the "six Es" – adding "ether" and "energy" to the four above. His "ether" refers to technology and the digital workspace, which may be measured through feedback or outages or even network usage. Usher's "energy" relates to the motivation and vitality of the workforce, rather than the conventional perception of energy in a building context, which may perhaps be measured using individual and team performance metrics.

The Vitruvian principles of architecture, introduced by the Roman architect and engineer Marcus Vitruvius Pollio, state that a building must exhibit the qualities of *firmitas*, *utilitas* and *venustas*, that is be solid, useful and beautiful. The principles date back to 30–15 BC but are still referred to and adhered to by modern architects. So, a POE may need to capture these three qualities through a mixture of occupant feedback and more technical engineering methods. The design quality indicator (DQI) assessment method, initiated by the Construction Industry Council in 1999 but still active, is based on Vitruvian principles. The design quality of a building is recorded at every stage of the construction process according to build quality and durability (*firmitas*), functionality and usefulness (*utilitas*), and impact and beauty (*venustas*).

Levels of POE

In their seminal book *Post Occupancy Evaluation*, Preiser, Rabinowitz and White (1988) introduced the notion of different levels of POE requiring increasing effort to complete. The Higher Education Funding Council for England (HEFCE, 2006) expanded on this concept to define three levels of POE in its guide to evaluating higher education institutions.

- **Indicative** – A snapshot of the strengths and weaknesses of the building project, including interviews with those most familiar with the building, accompanied by a walkthrough; a short, simple questionnaire might also be circulated to the occupants.
- **Investigative** – A more thorough evaluation undertaken to ascertain how a building complies with pre-set criteria, typically consisting of an occupant feedback survey backed up by focus group reviews and interviews.
- **Diagnostic** – A very thorough analysis which links objective measures of the physical environment to the subjective occupant perception of the building performance, and this may include measuring lighting, temperature, indoor air quality (IAQ), noise and energy consumption.

RIBA (2017) suggested that "POE can take on various shapes and sizes, from light touch to more intensive, and each POE should be tailored to individual project contexts to measure what is defined as important to the client and project team", while the *RIBA Plan of Work* (2020a) proposed the following three levels which, despite different terminology, overlay with the three original levels listed above.

- **Light touch** – Simple but meaningful rapid evaluation undertaken post-occupancy, before the building contract is concluded.
- **Diagnostic** – The light touch POE might identify the need for a more detailed evaluation.
- **Detailed (forensic)** – Investigations to identify and, where possible, resolve any significant and persistent performance issues.

The recently published British standard, BS 40101 Building Performance Evaluation of Occupied and Operational Buildings (BSI, 2022), clearly refers to BPE but also includes levels relevant to POE. The levels of evaluation range from preliminary investigation to light BPE to standard BPE and investigative BPE. All four levels include collating data on occupant/user experience, whereas the higher levels also include other metrics such as energy use, water consumption and internal environment monitoring (such as temperature, humidity and level of carbon dioxide (CO_2). The standard covers sustainability and tests whether buildings are performing technically as modelled and predicted, as well as collating occupant feedback.

Deuble and de Dear (2014) noted that "POEs should not only involve feedback from the building users, but also include the use of instrumental data, such as the measurement of indoor environmental quality (IEQ) indicators". They went on to say that occupant feedback may be affected by factors outside the remit of the project and that occupants are often asked to comment on past periods of time spent in the building (for example, questions about seasonal thermal comfort). Loftness et al. (2018) agree and propose that "An integrated approach to building performance evaluation mandates that POE subjective tools be matched by metrics (POE + M)". The metrics they refer to include measurements of air temperature, relative humidity, the levels of carbon dioxide (CO_2) and volatile organic compounds (VOCs),[1] air velocity, illuminance and other environmental conditions. I totally agree that a more detailed POE should include a mix of occupant feedback, indoor environmental quality monitoring and possibly other metrics, but the "M" is implicit, and the "POE + M" tag is therefore redundant, in my opinion.

Furthermore, while indoor environment measurements supplement occupant feedback, quite often, the two sets of measurements are weakly correlated. As I have explained elsewhere (Oseland, 2022), this may be due to "psychophysics": that is, the way humans respond to the physical environment is affected by both physiology and psychology, such that different people perceive the same environment differently. As such, researchers are interested in thermal sensation rather than temperature and noise rather than sound levels, and so on, and these factors can only be understood by collating occupant feedback. For this reason, it is better to start with reviewing feedback and then introduce more technical measurements if specific issues are highlighted. As the lead developers of *BS 40101 Building Performance Evaluation of Occupied and Operational Buildings* stated: "Occupants are the best (uncalibrated) building performance sensors we have" (Mashford and Gill, 2022).

Elements of a POE may also form part of a research study, whether longitudinal over many years or pre and post some form of research intervention or across a selection of buildings. Therefore, I have included POE methods that tend to be used more in research than in a standard POE, but fundamentally they can help us understand the impact of the building on its occupants.

Subjective versus objective

Social scientists differentiate between subjective and objective measures. In essence, "subjective" may be thought of as opinion and "objective" as fact. Occupant feedback methods are considered subjective because they seek personal views, attitudes and perceptions. Of course, some questions may be objective, such as "Are the lights switched on?" but most questionnaires use rating scales to collate a personal opinion, like "How do you rate the quality of the lighting?" Technical measurements, such a desk illuminance levels, are considered objective because the data is produced by an independent and observable measure.

Social scientists also distinguish between quantitative and qualitative data. Quantitative data is numbers-based and measurable, whereas qualitative data is interpretation-based and descriptive. For example, indoor environmental monitoring that produces a series of recorded measurements is quantitative, whereas interviews and focus groups mostly produce qualitative data. Interestingly, questionnaires can produce both quantitative data using rating scales or qualitative data through open-ended text questions. However, text (and spoken) responses can be quantified through, for example, content analysis where key words, phrases or themes are counted.

Components of a POE

What is included as part of a POE depends on the KPIs relevant to an organisation, the level of detail required and the preferred methods of those bodies promoting POE. And, of course, the POE components are also dependent on the resources (time, cost, expertise) available to conduct them.

Hadjri and Crozier (2009) reported in their literature review that "POE has progressed from a one-dimensional feedback process to a multidimensional process" with a portfolio of techniques. A mix of subjective versus objective, and quantitative versus qualitative, techniques are usually implemented. The key is to determine and use the most appropriate method for the project. They concluded that "in undertaking a POE, it is necessary to focus on the most relevant issues, rather than to attempt to analyse everything and face an overload of data" (Hadjri and Crozier, 2009).

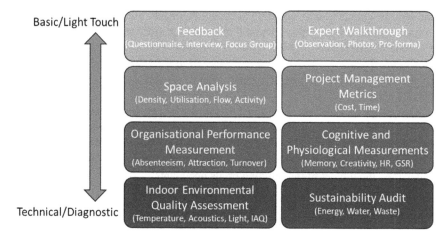

Figure 1.4 Core components of a POE

The following list is a short summary of the more common core POE techniques (illustrated in Figure 1.4), but the list is by no means exhaustive. The techniques are expanded on in Chapters 5, 7 and 8.

- **Feedback** – Most POEs involve obtaining feedback from a sample of the building occupants or anyone that may be impacted by the project. Building user feedback may be collated using questionnaires, interviews or workshops/focus groups, which may be carried out online or in person. The feedback may target a range of different stakeholders, including visitors and the project team. The feedback ascertains how well the project is supporting the occupants' comfort, wellbeing and performance and the quality of the project.
- **Expert walkthrough** – Most POEs include an expert walkthrough and review. This may be carried out by an independent professional and possibly conducted jointly with members of the design or facilities management team, who observe the building in operation. The walkthrough helps verify and explain the findings from the other POE techniques. The experts involved are also likely to pick up on technical design issues not observed by the occupants and are more able to judge the design quality. Photos may be used to support the observations.
- **Space analysis** – Real estate professionals are more likely to be interested in the functionality of the occupied space and how efficiently it is planned and used. A series of standard space planning metrics are often reported, such as workspace density and a breakdown of the ratio of open-plan desk space to other spaces. Commentary may also be made on the layout of the space and range of facilities provided. How the space is used is also of interest to those conducting a POE. Utilisation studies, sometimes referred to as either space or time utilisation surveys, may be conducted to find out how often the space (desk, private

office, meeting room etc.) is used related to work and occupancy patterns. If the study is carried out using observers, then they may also collate information on the activities carried out by the occupants during the day. Observers may also record the flow of movement through the building, sometimes carried out in combination with space syntax analysis. Ethnography is less used in POE but is an established methodology used in anthropology and involves the observer studying the behaviour of those in a particular social situation.

- **Project management metrics** – The most common and fundamental metric for assessing the performance of a building project is cost. The cost metric used is usually the capital expenditure (often referred to as CapEx) for the project, but operational expenditure (OpEx) may be included. Project managers will undoubtedly use cost as part of their evaluation as well as the time to completion – "on time, on budget" is a common mantra. However, they may also use a POE to explore other factors affecting project delivery, such as the project team briefing process, procurement process, sharing of information, decision making and resolution of issues. Cost (£) could be combined with measures of quality (Q) to calculate and compare value ($V = Q \div £$).
- **Organisational performance measurement** – Where practical, a POE may include human resources (people-related) performance metrics such as attrition/turnover, attraction/recruitment, absenteeism, staff utilisation, wellbeing and training costs. Additional business metrics related to individual, team and organisational performance may also be collated (for example, sales, deliverables, income/turnover and profit/margin). Organisational performance measurement includes health and safety (H&S) metrics, such as accident incident rates, absenteeism and lost days (downtime). H&S is important to all companies but key to some, for example, those in manufacturing or the oil industry. Due to their sensitivity, organisational performance metrics can be difficult to acquire and are often confounded by other factors outside the remit of the project, so they are rarely used as part of a POE. However, self-assessed performance and wellbeing could be included as part of the occupant feedback survey.
- **Indoor environmental quality assessment** – More detailed technical POEs may require measurement and monitoring of the indoor environmental conditions such as temperature, ventilation rates, IAQ, acoustics and light. Such measurements are complex, with each parameter involving different variables and techniques.
- **Sustainability audit** – An increasing focus on net zero carbon and how well a building is performing compared with predictions, along with the uptake of building certificates such as BREEAM, LEED and WELL, has resulted in an increase in environmental monitoring. Energy/utility costs (electricity, gas and oil), water consumption and generated waste are common metrics.
- **Cognitive and physiological measurements** – These measurements are quite niche and tend to be used only as part of a wider research study rather than a standard, or even detailed, POE. Nevertheless, some investigators (see Gillen, 2015) do use online and app-based cognitive performance tasks as part of a POE, including memory recall, attention, language, abstract thinking and mathematical tests.

With advances in wearable technology, future POE studies may include physiological measurements, such as heart rate, galvanic skin response and blood pressure.

In several previous papers and my previous book, *Beyond the Workplace Zoo*, I explained the importance of measuring the success of a workplace or building the business case for a project based on effectiveness, not just efficiency or, to be more precise, based on value and productivity rather than cost alone. I consider productivity to be the ratio of output to input where, in the workplace industry, the output relates to human resources and business performance metrics, and the input relates to the cost of providing the workspace. The POE metrics above overlap with my views on calculating productivity, and in that respect reviewing productivity can be considered a fundamental part of workplace POEs.

The future of POE

Space utilisation, such as the proportion of occupied desks, may be included as part of a POE. Traditionally utilisation was recorded using observers who visited the spaces hourly over a two-week period; it is now more common for sensors to be used to monitor occupancy continuously over several months or more. Some of the larger corporate occupiers have even deployed sensors to monitor office space in real-time to help ensure maximum utilisation. Similarly, sensors that continuously monitor indoor environmental conditions are becoming popular. For example, The Edge, the Amsterdam headquarters of global consultancy Deloitte, was fitted throughout with sensors to monitor and adjust the environmental conditions in terms of both comfort and sustainability (Randall, 2015). The development and reduced cost of this technology has resulted in a trend for real-time continuous monitoring rather than one-off surveys, such as pre- and post-project measurements. But while occupancy and environmental sensors tend to be fixed (for example, under-desk or overhead), the trend of wearable technology offers an alternative, especially for monitoring space usage or perhaps for recording physiological and health metrics.

Ongoing, instant and real-time feedback is also becoming more prevalent: consider the "smiley face" pushbuttons at airports and in public toilets. Organisations such as the taxi firm Uber ping a message after each ride requesting their customers to rate their journey, as do airlines and train companies. Similarly, hotel booking platforms (websites or apps) email their customers after each stay with a short feedback survey. In these examples, the request for feedback is in response to a recent action or event rather than issued at a set time, such as an annual survey. The feedback requested is also usually quite specific and short rather than a lengthy post-project survey.

Going forward, POE information may be collated in real-time and in short bursts in response to specific building interventions. For example, feedback may be sought after an occupant visits a newly refurbished space (restaurant, meeting room, reception etc.) for the first time. QR codes are becoming more prevalent as a means to assess spaces: the visitor simply aims their phone camera at the QR code and is directed to an online questionnaire. This requires initiation by the visitor to the space, but occupant sensors, facial recognition and other tracking devices could determine when and who to request feedback from (for example, after they leave a space that is being evaluated). The data could be collected and combined over a longer period rather than in a one-off online survey. Another opportunity is to link real-time feedback on the environmental conditions to the building management system (BMS) so that unsatisfactory conditions can be automatically adjusted.

The means by which occupant feedback is obtained may also change. Most of the online survey platforms are now compatible with mobile devices, allowing participants to respond more easily while on the move rather than needing to be at their laptops. Some survey platforms now allow images to be added as prompts or as responses. Software, such as Cognito UX by The Curve AI,[2] uses photos, graphics and gamification to provide a highly visual and interactive means of capturing feedback.

Back in 2005, the "Bop! Project" (Wilson, 2006) explored alternative means of collating occupant feedback using a variety of ubiquitous wireless sensors and quirky interactive devices. My favourite of these techniques was two pressure-sensitive mats, one with "yes" and one with "no" printed on them. A different question was posted above the mats each day, and people voted as they entered or left the building simply by stepping on their chosen mat. Other methods included pushbuttons and dials dotted around the space with different questions. Although data gathered in this way is less detailed, it is quick, accessible, in real-time, and likely to improve response rates. This approach is more likely to appeal to those who are not fans of lengthy online questionnaires and are looking for more proactive and ongoing feedback.

Artificial intelligence (AI) and voice and face recognition are among the more recent advances in technology that may be adopted in POEs. Some survey platforms offer AI based sentiment analysis as part of their offering. The UK start-up Audiem[3] has developed an AI platform for interrogating large quantities of data, particularly open-ended questions with lots of free-form text or comments made in chat room threads (like the Slack app). Free-form inputs are often avoided, given the resources required to absorb them and make sense of the feedback gathered. However, AI platforms like that of Audiem provide an opportunity to collate and interpret stories and comments, including verbal as well as written feedback. Voice recognition could be used to capture occupant feedback, rather than through typed comments, forms or pushbuttons, and then subsequently analysed using AI platforms. Disparate data sets from a range of media sources could be collated and combined.

Notes

1 There are thousands of VOCs, so the total volatile organic compounds (TVOC) is usually measured as it is easier and less expensive than measuring individual VOCs.
2 www.thecurveai.com/services-3.
3 www.audiem.io/

2 Why (and why not) POE?

Having explained that POE is a means of evaluating the performance of occupied buildings, using occupant feedback and other techniques, my next task is to discuss the important question of why a POE in some form should be initiated.

Benefits and purpose of POE

Sir John Egan (1998) told us the importance of seeking occupant feedback on building projects, likening it to the way retailers seek customers' views on their products and services. Failing to find out what people think of the buildings they use, particularly offices, is a lost opportunity on many fronts. For instance, gathering end-user feedback and acting upon it can:

- determine whether the building adequately supports the vision and activities of the occupying organisation,
- quickly identify building faults or problems (including design, layout, facilities and indoor environmental quality) that can be resolved to everyone's benefit – often at a considerable financial saving to the organisation,
- improve wellbeing and performance, and therefore profitability, because the space provides a more comfortable environment and better supports the occupants' activities,
- enhance staff satisfaction and morale because the occupants' opinions are being taken seriously, and their needs will be better fulfilled,
- raise the profile of the corporate real estate and facilities management teams by proactively seeking customer feedback and measuring their satisfaction.

Of course, occupant feedback is only one of the techniques regularly used in POEs. The POE may also include more technical evaluations, such as indoor environment measurements, space analysis, cost analysis, energy audits etc. But triangulating the results of a combination of subjective and more objective metrics results in a more robust and valid POE.

For those who are involved in creating, acquiring, managing or occupying office buildings, the key benefit of a POE is that it provides an understanding of whether the design, planning and operation of the building support the activities of the

DOI: 10.1201/9781003350798-2

occupying organisation. A good POE will identify whether the occupants' (and the project team's) requirements and brief were realised and if the anticipated benefits of planned improvements were actually achieved. Similarly, POE is increasingly used to check that energy use and indoor environmental conditions match those modelled and planned. A POE will also unearth lessons learned and provide guidance for future projects.

In major fit-out or refurbishment projects occupant feedback and other POE metrics can be collated pre-project implementation as input to the brief and to inform the design. A POE can also be used to advise the project team of potential improvements to the project and any lessons learned that may be used to enhance the quality of the existing project or future similar projects.

I am reminded of the story of the *Three Little Pigs* and how they improved their houses to withstand destruction by the wolf (see Figure 2.1). Many building projects are one-offs; indeed, Zimmerman and Martin (2001) argued that every building is, in effect, a prototype. Nevertheless, lessons learned from one building project can, and should, inform the next one. Interestingly, on a trip to Cuba, I noticed that the design of some buildings, particularly schools, was repeated across the island – they were rolling out a tried and tested design.

Furthermore, regular occupant feedback and monitoring between projects help manage building performance and fine-tune a facility. Proactively seeking feedback and continuous measurement can minimise the risk of any building or facilities problems becoming serious or, as Way and Bordass (2005) noted, it will "help prevent minor problems developing into longer-term chronic irritants". Thus, in addition to assessing building projects, POE equally applies to the evaluation of projects such as new services and processes, and to regular business-as-usual building evaluations. Proactive POE can also inform how well the space is utilised and corresponding opportunities for space and associated cost, savings or readdressing the balance of what different spaces (work settings) are required.

More recently, POE has been used to compare the performance of the building in use to the performance modelled and predicted. In particular, the energy performance of the occupied building is measured and compared. This rationale for POE has become more relevant as companies strive towards their net zero carbon

Figure 2.1 Lessons learned to improve building design

Source: Clara Doty Bates, 1885, Wikimedia Commons

targets. Water consumption and waste may also be monitored and compared with the levels predicted at the design stage of a project.

Ultimately, the long-term benefits of conducting a POE are improved occupant comfort, wellbeing, health and performance, along with more sustainable and cost-effective buildings. As RIBA (2020b) summarised, "POE makes buildings greener, cheaper and more productive", pointing out that POE offers a range of benefits: "From lower energy bills to reductions in carbon emissions, a culture of continuous improvement can deliver huge returns on investment". RIBA stated that POE could make buildings greener and healthier, reveal how a building is used compared with its designers' intentions, reduce operational costs, allow for continuous improvement, reduce delays and overspending on future projects, and increase user satisfaction.

In summary, the key purposes of conducting a POE as part of a building project include the following.

- **Feedback and feedforward** – Informing the project team of the successes and weaknesses of the project will help improve future projects and support continuous improvement. As Heath et al. (2019) put it, "you get the chance to find out: What has worked well and should be built upon . . . What didn't work as well as hoped . . . What should be done differently to improve current or future projects". There may be minor issues with the current project that can be fixed relatively quickly and easily, particularly if related to project elements under warranty. Feedback should cover all aspects of the building, such as design, layout and facilities provided, space use, indoor environmental quality and sustainability.
- **Measure project success** – Most project leaders need to justify project expenditure and show that what was required and commissioned was delivered in line with the level of investment, achieving the required quality and value. This post-project analysis is sometimes referred to as "benefits realisation". A robust cost-benefit analysis will include less-tangible as well as tangible benefits. A POE can highlight these: as well as possibly showing space, energy and cost savings, it can be used for determining levels of occupant comfort, wellbeing and performance and support of occupant activities and organisational goals.
- **Actual versus predicted performance** – With more emphasis on building sustainability, POE is being increasingly used to compare the in-use building performance with that predicted using design and planning models. Indoor environment measurements of temperature, lighting and so on may be employed along with energy, water and waste audits.

If a pre-project evaluation is conducted along with a post-project one, then the following can also be provided by the POE process.

- **Inform the design** – Pre-project feedback helps determine the occupants' requirements and ensures the majority opinion is captured rather than minority views. The priorities for the project can also be determined and used to focus expenditure, especially when there is a limited budget. A pre-project POE that includes feedback and space analysis can be used to determine and fine-tune the actual space required, potentially saving space and, in turn, property and energy costs. Other collated POE elements, such as occupancy levels and operational costs, all inform the design brief and project success factors.

- **Baseline for measurement** – Taking measurements and conducting analysis prior to the project and after completion makes it possible to compare the project with the original building conditions. This is a particularly good approach for reporting highly subjective factors such as changes in self-assessed wellbeing and performance. Pre- and post-project evaluations also provide a more robust comparison than using third-party (external) benchmark data.
- **Change management and communication** – Collating occupant feedback at the start of a project and prior to, say, a building fit-out or refurbishment improves occupants' involvement by allowing them to present their ideas and contribute directly to the project. A survey administered to all occupants and key stakeholders is a convenient way of making everyone aware of the project. The POE can be linked to broader in-house communication programmes and change initiatives.

Before deciding on the techniques to be used for the POE, it is important to agree on the objectives of the POE. Those objectives should be SMART ones: Specific, Measurable, Achievable, Relevant and Time-bound. For example, if one objective is to measure the impact of the building on occupant wellbeing, then it is crucial to agree on how wellbeing will be measured. At the start of a building project, I sometimes ask the members of the project team to state their top objectives and ask them for SMART ones. The objectives proposed by team members usually vary (see Figure 2.2), so a vote is then made to determine the most relevant ones. Those agreed objectives can be revisited during the project programme and in a post-project evaluation.

Figure 2.2 Different views of project success

Source: Aukett Swanke

Further justification and advocates of POE

Several professional bodies, certification systems and governments call for POEs to be integrated into the design process and a building's life cycle. This section lists some of the notable influencing bodies that are raising awareness and making POE more accessible.

UK government and non-governmental bodies

The Office of Government Commerce (OGC), formerly part of HM Treasury, encouraged the uptake of POE in the UK when it introduced annual evaluations of all central government buildings, including post-implementation reviews, as part of the *OGC Gateway Process Review 5*. The OGC was closed in 2011, but its successors, the Government Property Unit and the Office of Government Property, have maintained the process and in 2021 published a revised and updated document, *Gate 5 Review: Operations Review and Benefits Realisation*. The Infrastructure and Projects Authority (IPA) noted that the "process is designed to provide a realistic view on a programme and project's ability to deliver agreed outcomes to: time; cost; benefits; and quality" (IPA, 2021). The review process includes a series of questions and requires evidence under three sections (core, infrastructure and transformation) to help determine whether "the benefits set out in the business case are being achieved and that the operational service (or facility) is running smoothly and the agreed strategic outcomes are being met". Gathering the required evidence involves a range of techniques, including stakeholder feedback and sustainability performance metrics, all documented in a benefits realisation plan.

In 2013, the Cabinet Office introduced the *Government Soft Landings* (GSL) policy alongside BIM. The GSL emerged from the Soft Landings approach conceived by Mark Way (see Way and Bordass, 2005) and refined by the Building Services Research and Information Association (BSRIA), which promoted soft landings long before the UK government launched the GSL. Since 2016, GSL has been a requirement for all UK centrally procured public sector projects, and in 2019 it was updated by the UK BIM Framework to include the British and international standards *BS 8536* and *ISO 19650* (see Figure 2.3). As with the original GSL, the latest version calls for post-project POE to be conducted over a three-year aftercare period to measure functionality and effectiveness along with environmental performance and cost performance.

The HEFCE, which distributed funding to universities and colleges in the UK between 1992 and 2018, required a POE to be completed in order to satisfy that the funding was used appropriately. To assist with the evaluation, HEFCE and the Association of University Directors of Estates introduced a best practice guide to POE (HEFCE/AUDE, 2006). Following on from HEFCE, the Skills Funding Agency (2014) also required a POE to be completed by higher education institutions that it provided with capital funding, whereas the newly formed Education and Skills Funding Agency places less emphasis on POE.

Figure 2.3 Best practice guides identifying the need for POE

Other countries

A few government departments in other countries also encourage POEs to be conducted. Carthey (2006) discussed the *New South Wales Standard POE Methodology* for Australian health projects, noting that the New South Wales (NSW) Treasury had published its *10-Step Procurement Process – Construction* in 2004, finishing with "Step 10 – Evaluation". In the same year, the NSW government published guidance on post-implementation review, which "collects and utilises knowledge learned throughout a project to optimise the delivery and outputs of future projects", and the NSW Department of Health made post-implementation review a mandatory requirement in their construction projects. Similarly, the Queensland Department of Housing and Public Works published a guideline on POE in 2017, which refers to the Project Assessment Framework administered by the Queensland Treasury. It "promotes benefits realisation reviews which focus on ensuring that a project or a program delivers the anticipated benefits and value for money documented in its business case . . . The POE has a key role in the process efficiency improvement and in contributing to improved service delivery outcomes".

While most government departments do not mandate that a POE must be conducted, they may be encouraging uptake by requiring their own buildings to meet specific certification requirements, such as *Leadership in Energy and Environmental Design* (LEED) or the *WELL Building Standard*. For example, in the US, the General Services Administration commissioned evaluations for 22 of its sustainably designed (green) buildings in order to gain LEED or ENERGY STAR certification (Fowler et al., 2011).

Nevertheless, although the case studies presented later in this book show that POEs are conducted globally, it appears that the UK, the US and parts of Australia are the early adopters because POEs are not a priority for most other governments, and it is rare for POEs to be a mandatory requirement.

British and international standards

British Standard BS 8536: Briefing for Design and Construction, referred to in the UK's GSL policy, applies to all new building projects and major refurbishments and is for use by architects and all construction professionals. However, it is important to note that a British standard is considered best practice rather than a mandatory requirement. *BS 8536* includes the briefing requirements for BIM, POE and Soft Landings. The international standard *ISO 19650* defines a common international framework for the effective production and management of information across the full life cycle of a built asset. There are six parts to *ISO 19650* with part 4, updated in 2022, focusing on using BIM Level 2 and superseding an earlier British standard *PAS 1192*. The UK was an early adopter of BIM, but the creation of a relevant ISO standard is indicative of growing global interest and potential wider adoption of Soft Landings, BIM and POE.

 BS 8536 encourages POEs to be carried out, whereas the most recent British standard provides guidance on how to conduct a BPE, which includes a POE. *BS 40101: Building Performance Evaluation of Occupied and Operational Buildings* provides guidance on the ongoing evaluation of a building. When it launched *BS 40101,* the British Standards Institution (BSI, 2022) explained that it "includes data gathering relevant to the BPE level pursued, through documentation review, test and measurement, monitoring, experiential feedback and observations along with a comparison against comparator case parameters". The new standard is a well thought-out, detailed and laudable document. I have a slight concern that the level of detail may discourage rather than encourage the uptake of POE, but that is the nature of standards covering all eventualities.

Building certifications

The uptake of building certification has gradually been increasing worldwide, particularly for those related to sustainability and wellbeing.

 The *Building Research Establishment Environmental Assessment Method,* better known as BREEAM, was the first method of assessing and certifying the sustainability of buildings. All types of BREEAM, including in-use (BRE, 2016), new construction and refurbishment, award a credit for conducting a POE as part of the assessment category "Management 05 Aftercare". The category calls for "a review of the design intent and construction process" and "feedback from a wide range of building users including facilities management on the design and environmental conditions of the building". The BREEAM POE also requires a review of sustainability performance, including energy consumption, water consumption and performance of any sustainable features or technologies.

 The LEED programme is another sustainability certification that is used worldwide, but it was developed in the US. LEED did not originally require a POE to be conducted, but it now requires the reporting of water and energy consumption during the 5-years after the occupancy of the building (US Green

Building Council, 2019). This monitoring is used for determining whether the actual performance of buildings that pursue a LEED certificate matches the anticipated performance.

The *ENERGY STAR* certification, managed by the US Environmental Protection Agency, requires a site visit to verify performance, and this includes measurements to assess indoor environmental quality along with observations of occupants looking for signs of discomfort.

The *National Australian Built Environment Rating System* (NABERS) is "a voluntary performance-based rating system that measures an existing building's overall environmental performance during operation". Initially developed in Australia and managed by the government of NSW, NABERS was launched in the UK in 2020 and is gradually being used elsewhere beyond Australia. Although NABERS is a certification system, the approach may also be considered a POE because an occupant satisfaction survey is required to satisfy the qualitative assessment of how a building is performing from the perspective of its occupants and it also includes indoor environment measurements and space analysis.

The *Ska Rating* environmental assessment method and benchmark are arguably less well known than those described above. However, the scheme has two relevant points of interest: (1) it includes an optional occupancy review to determine "how well the fit-out has performed in use against its original brief"; and (2) the Royal Institute of Chartered Surveyors (RICS) now manages it and hopefully encourages its membership, mostly consisting of quantity surveyors, cost consultants and project managers, to implement POEs.

The *WELL Building Standard* "is a performance-based system for measuring, certifying, and monitoring features of the built environment that impact human health and wellbeing, through air, water, nourishment, light, fitness, comfort and mind" (IWBI, 2016). The certification includes concepts and features and in WELL v2 the "community" concept incudes "Feature C04 Occupant Survey" and "Feature C05 Enhanced Occupant Survey". WELL v2 recognises and recommends a list of 13 approved third-party surveys,[1] including the CBE Occupant Survey, the Leesman Index, Building Use Studies (BUS) occupant feedback and wellbeing survey, Sustainable and Healthy Environments (SHE) and PeopleLOOK by Baker Stuart.

Other wellbeing certification systems are available, such as Fitwel, originally developed by the US Center for Disease Control and Prevention and the General Services Administration. Fitwel is a scorecard consisting of over 60 strategies that affect seven "health impact categories". To certify, Fitwel requires documented evidence but not necessarily a POE. However, additional points can be gained by adopting its *Enhanced Indoor Air Quality Testing Policy* (2021) which does require a POE and measurement of IAQ metrics such as the level of CO_2, VOCs and relative humidity.

There are undoubtedly other sustainability and wellbeing certification systems in use around the world, but those listed above highlight how such certifications encourage the uptake of occupant feedback and other elements of POE.

Professional bodies

The original RIBA "Plan of Work", dating back to 1963, included "Stage M Feedback" but at some point it was dropped until the publication of the then newly formatted *RIBA Plan of Work 2013* with its seven work stages. The most recent version (RIBA, 2020a) heavily features the use of POE, particularly in stages 6 and 7, but also with some elements at stages 0 to 5 (Figure 2.4).

"Stage 6. Handover" states that "making sure that the building is performing as anticipated after occupation requires a light touch post-occupancy evaluation to be undertaken"; furthermore, "the project team will be interested in the feedback from a light touch post-occupancy evaluation, conducted once any seasonal commissioning has been completed, so they can understand how the building is performing and whether the building and its systems are being used as planned". POE is key to "Stage 7. Use" which recommends that practitioners "complete the aftercare activities, such as detailed post-occupancy evaluation", which "are commissioned to determine how the building is performing in use to help fine-tune the building and inform future projects".

Each stage of the plan highlights a list of strategies and tasks, including "sustainability" and "plan for use". Both these strategies at all stages include POE, such as reviewing previous POEs and identifying the lessons learned relevant to the current project or planning the requirements to be checked later during the POE.

The significance of the *RIBA Plan of Work* to POE is that it is the definitive framework, or the design and process management tool, for UK architects and the wider construction industry to use on building projects. The framework is also used as best practice by architects based in a number of other countries. As the *RIBA Plan of Work* includes POE as part of the design and delivery process for building projects, it is key to the construction industry's adoption of POE.

RIBA has continued to champion the importance and uptake of POE through its membership, lobbying and a series of publications (RIBA, 2016, 2017, 2020b), stating in particular that "Regular evaluation is standard in the most innovative businesses and, we argue, should be standard in architecture" (RIBA, 2017).

Figure 2.4 RIBA Plan of Work

Cooper (2001) suggested there was a lull in POEs in the 1990s because they had dropped from the curriculum of British schools of architecture, but POE was later revived because it was adopted by the newly emerging discipline of facilities management. My first consultancy role was conducting POEs for a facilities management company in the late 1990s. That business, like the wider facilities management community, recognised the usefulness of conducting regular POEs to help determine the satisfaction of their customers (i.e. the building occupants) and used the feedback to improve their service offering. At the time, I worked with Ian Fielder, who later became the Chief Executive of the British Institute of Facilities Managers (BIFM, now rebranded IWFM), and he remarked that "BIFM considers POE important to facilities managers as it provides direct feedback on our ability to operate buildings to a high standard . . . POE must be part of the design process and budgeted for to ensure it becomes a business-asusual activity".

In 2016, BIFM revised its *Operational Readiness Guide*, which offers best practice guidance on creating a building that has the end-user in mind, as shown in Figure 2.5 with other key guides. The guide recognises the growing requirement for BIM, the need for monitoring post-occupancy energy performance and references Stage 7 of the *RIBA Plan of Work*, "providing the opportunity for new post-occupancy services that will help to ensure that a building is running as intended and effectively for the user". A more recent publication, *Employer's Information Requirements* (IWFM, 2017), provides facilities managers with guidance on projects requiring BIM and also references RIBA's call for POE.

It is less clear to me whether the International Facility Management Association (IFMA) actively endorses POE, but it does acknowledge and support POE case studies. For example, occupant feedback was collated as part of the case studies in its *Workplace Amenities Strategies Research Report #36* (IFMA, 2012). I have also seen case studies, including POE results, presented at the IFMA conferences. Back in the mid-1990s, IFMA piloted a survey for obtaining academic facility performance feedback.

I mentioned in the preface that this book is a revision of the BCO *Guide to Post-Occupancy Evaluation*. The BCO commissioned that guide as a direct response to the Egan Report of 1998. The BCO's mission is to research, develop and communicate

Figure 2.5 Call for POE by professional bodies

best practice across the office sector, and the original 2007 guide stated that "Conducting and sharing POEs of our workspaces supports this objective and this guide is a testament to BCO's belief in the value of POE". Since then, the BCO has provided further endorsement of POE, including an annexe to the original guide in 2011 with a focus on BPE and sustainability.

In the late 1990s, the monthly magazine of the Chartered Institution of Building Services Engineers (CIBSE), the *CIBSE Journal*, published a series of POE studies. The methodology was termed PROBE, or Post-Occupancy Review of Buildings and their Engineering. There were two waves of studies with 16 buildings evaluated. PROBE included the BUS occupant feedback survey, a site visit and building inspection, an energy assessment and, in later studies, water consumption and pressure tests for air leakage.

CIBSE continues to promote POE, particularly in the area of energy consumption, through its series of technical memorandums and case studies. For example, *TM22 Energy Assessment and Reporting Methodology* (2006) is a method for assessing the energy performance of an occupied building based on metered energy use. The related *TM23 Testing Buildings for Air Leakage* (2022) describes the fan pressurisation and low-pressure pulse methods for measuring air tightness. CIBSE has also produced occupant questionnaires: see *TM62 Operational performance: Surveying occupant satisfaction* (Bunn, 2020) and one for schools featured in *TM57 Integrated School Design* (CIBSE, 2015b).

BSRIA, a UK-based independent building research and consultancy, launched the *Soft Landings Framework* in 2009, based on the earlier research of Mark Way (see Way and Bordass, 2005), and revised it in 2018 to align with the *RIBA Plan of Work*. BSRIA's framework includes extended aftercare and POE, and its guide *Building Performance Evaluation in Non-Domestic Buildings* (Agha-Hossein, Birchall and Vatal, 2015) includes three main POE activities: occupant satisfaction, indoor environmental quality and operational performance. The guide recommends using the BUS and Leesman surveys for measuring occupant satisfaction. BSRIA also provides POE and BPE as a service, including indoor environmental quality monitoring, energy analysis, forensic walkthrough, airtightness testing and thermal imaging.

The BRE is another UK-based independent research body that has been active in the field of POE. When I worked there back in the 1990s, I developed a POE methodology with my colleagues in the Human Factors section. Paul Bartlett and I developed an occupant feedback survey which we published in *Improving Office Productivity* (Oseland and Bartlett, 1999). The intention was for the survey to be made freely available to all, and it was adopted by several organisations that used it to evaluate numerous buildings. For example, the Occupiers Property Databank used a version of it to evaluate the central government property portfolio. BRE still promotes POE and offers a range of related services, including its Design Quality Method (DQM), monitoring of environmental conditions and sustainability audits.

Regarding promoting the uptake of POE, the Leesman Index survey is worthy of mention. There are many standardised occupant feedback surveys on offer, but since its launch in 2010, the Leesman is the one that has grown most rapidly. The Leesman team have now collated over one million responses from more than 6,500 workplaces in over 100 different countries, thus allowing benchmarking across countries and industry sectors. Not only is the database of user experience growing, but the founder Tim Oldman is a regular and passionate speaker on the importance of using feedback to create better workplaces and enhance employee experience.

While it is clear that UK organisations were early adopters and advocates of POE, the approach is promoted elsewhere around the world. For example, the World Green Building Council campaigns for improved wellbeing and performance in offices. Its 2014 report *Health, Wellbeing and Productivity in Offices*, provided a framework for collating evidence and data gained through POE.

The Environmental Design Research Association (EDRA) is an international membership organisation founded in the US in 1968. Its members include social scientists, architects and designers interested in the interrelationships between people and their surroundings. Many members conduct and promote POE, and the association has a POE/programming network. Similarly, the International Association of People-Environment Studies has a history of promoting POE to evaluate architectural projects.

Barriers to POE

The benefits detailed earlier in this chapter and the range of case studies in Chapter 10 clearly demonstrate the value of and need for POE. However, RIBA (2020b) recently reported that "Only 19% of practices in the UK offer clients a POE service . . . only 3% of practices always measure actual or anticipated operational energy through POE, and 50% never do". Watson (2020) noted that the *2020 AJ100 Survey* found similar results, with just 4% of architectural practices always conducting a POE and 48% evaluating occasionally. A North American survey by Skidmore, Owings & Merrill (Fairley, 2015) found that 62% of architect firms carried out POEs, but they were conducted on only 5% of their projects. So, there is some very gradual uptake, predominantly in the UK, but despite the outlined benefits, POE is not yet a routine or valued activity. There are four primary reasons for this lack of uptake, but each barrier is readily overcome (Figure 2.6).

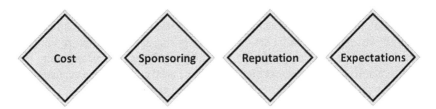

Figure 2.6 Four barriers to POE

Cost

In my experience, a core reason cited for failing to conduct a POE is cost. Most organisations will not have budgeted for a POE as part of their project and may simply not have the funds to finance one. The complexity and breadth of techniques used in POEs may be off-putting to some if they are unsure of what the POE entails and how much they will be invoiced for. For example, are they commissioning an occupant feedback survey and high-level review or a full-on technical appraisal with continuous environmental monitoring? This means POE experts need to explain the levels of POE and the associated components and agree on what type of POE is required and to be budgeted for.

Rather than cost *per se*, the barrier may be a perceived lack of "value", where value may be considered as a return on investment or simply "what is achieved for a certain cost". On project completion, the POE provides valuable feedback to the project sponsors or executive board because it provides measures of tangible factors, such as space and cost savings, and also the less-tangible factors, such as quality, comfort, wellbeing and worker performance.

Some organisations may occupy a portfolio of buildings, and carrying out POEs across all their buildings would make it possible to compare the value provided by the buildings rather than simply comparing their costs alone. The results of these POEs could be the foundations of a database for setting organisation-specific benchmarks. Furthermore, on a major project, an occupant feedback questionnaire could be used as part of the briefing process, and the results could be used to inform the design, focus expenditure and input to the business case for the project.

Questionnaires are a cost-effective means of obtaining widespread information at the pre-project briefing stage, and the same questionnaire can be re-used at the post-project stage. For example, one of my architectural colleagues told me there used to be a percentage allowance in projects for "art". Similarly, I recommend that a budget should be included for POE at the pre- and post-project stages. RIBA (2020b) reported that "The cost of POE is a very small percentage of overall building costs . . . undertaking POE adds an additional 0.1%–0.25%".

Another type of cost to the occupier is the time it takes their staff to participate in feedback (questionnaires, interviews, workshops) or the disruption caused by the more technical POE methods. Some occupants may suffer from survey fatigue, as feedback is ubiquitous (except in the building industry). If time is an issue, then feedback surveys can be kept light, as a few questions are better than none, and using online surveys means they can easily be completed on a mobile device in downtime, for example, when on a coffee break or travelling. Technical POEs are more intrusive but can be set up out of working hours and left running, and some sensors (for example, indoor environment and occupancy) are quite small and discreet. Existing documentation can also be used as part of the evaluation; for example, using the space plans to conduct a space analysis.

Sponsoring

Related to cost is the question of "who pays for the POE?". This is more to do with ownership – most projects have a sponsor – and the same is required for POE. The project sponsor usually advocates, commissions and pays for the POE. However, there is some reluctance to champion and pay for POEs among the occupier, design team and facilities manager because they each see their counterparts gaining the most benefit from the study.

Quite often, particularly with a one-off project, the occupier may consider the design team to gain the most from conducting a POE, such as feedback and lessons learned, so they are not inclined to pay for such an analysis. Whereas, in reality, the occupier benefits by knowing whether they (and their commissioned project team) met the project objectives, fulfilled the needs of their workforce, and delivered value. In most organisations today, there is an expectation that the project lead and design team will not only deliver value but prove it, sometimes referred to as benefits realisation.

The many benefits of conducting a POE need to be clearly explained to the occupier, but the designer does, of course, benefit hugely from the POE. When I worked for an architectural practice, we offered POE as a value-add service. My architectural colleagues not only found that their clients appreciated their interest, but quite often, they commissioned them to carry out more design work. The post-project evaluation was a small expense because an occupant survey was usually carried out as part of the briefing process, and a similar survey was used post-project.

The facilities management team also benefit from commissioning POEs. Surveys that are repeated, say, annually and which use the same questionnaire will cost less than one-off surveys and also provide a database of comparable results. If the survey helps identify preventative maintenance, then it could save on unnecessary future expense. Such a POE also provides useful feedback for out-sourced facilities management companies who are about to renew their contracts.

Reputation

Designers and architects may be concerned that the POE will raise issues with the design that may, in turn, affect their reputation: in particular, the exposure of the project team and the potential liability regarding any problems or defects identified through the evaluation for which they are accountable.

If the project is genuinely poor, then it needs to be flagged because the occupants, not just the reputation of the designer, will be affected indefinitely. However, the POE is usually carried out after the defect liability period and the commissioning phase. In practice, it is unlikely that the POE will reveal any unknown major defects; the POE is more to do with quantifying the quality of the workplace, determining how well the design benefits the occupying organisation, and picking up on any rectifiable criticisms identified by the end users. It relates more to how the building supports the occupying organisation, which depends on the brief, how the business uses the building and other factors outside the control of the designer. POE in the public domain is possible at a high level of scientific integrity without attracting litigation or technical disputation.

Furthermore, research has shown that the reverse is usually true, and the project team's involvement in POE demonstrates that the design team are seeking improvement by proactively trying to understand how their buildings work for the users. My experience is that a POE is usually greatly appreciated by the occupier and seen as a joint learning, rather than blaming, opportunity.

The above comments assume that the designer/architect will conduct their own POE of their project. However, it is preferable that it is conducted by a third-party unconnected with the project in order to provide an independent and unbiased view of the building's performance. Nevertheless, it is better that a POE is conducted by the designer than not at all.

Expectations

Another concern is that POE may lead to an expectation that further changes will be made after the project budget is spent. Again, my experience is that POE can identify behavioural changes with no additional capital costs or identify other quick wins with minimal costs. However, if a genuine problem is identified through the POE, then it ought to be resolved, especially if it is a problem that is believed to be having a detrimental effect on worker performance and incurring financial losses.

Facilities managers might argue that a POE will open the floodgates to more problems or claim that they have not received any complaints, and so do not require a POE. Alas, in some companies, the occupants may feel they have no easy mechanism for suggesting an improvement to their workplace, and generally, a complaint means an issue has become a serious problem – one that might have been avoided with more proactive surveying and building management. I worked for a facilities management company, and at our annual account holders conference I proposed that we conduct regular POEs of our clients' offices. One facilities manager raised his hand and objected, saying POE would "open a can of worms", but fortunately the Managing Director at the time sided with me saying we should have nothing to hide and should actively seek feedback and fix any issues raised.

The advantage of POE is that systematically gathered occupant feedback shows representative views rather than anecdotal comments, which are often from a more vocal minority and cannot be reliably attributed to the whole of the organisation. A POE is more effective when the results are communicated and some action is taken in response to the survey, but such actions do not have to be expensive and sometimes good ideas are identified that actually lead to financial savings. It is important to share the findings of the POE and if any issues are uncovered then explain how they may be resolved in the future. Also, it is important to provide a balanced POE reporting the successes and positive benefits of the project and phrasing criticism as recommendations.

Related to expectations is the type of questions asked. Some organisations, particularly their HR departments, may be concerned that sensitive or organisation-related questions (about management, reward, etc.) are being asked. However, sensitive questions can be avoided, especially when using tried and tested questionnaires, and questions asked can focus on the physical features and design of the workplace.

Sharing POEs

In my opinion, the benefits of conducting a POE far outweigh the barriers. A POE is essential to improving the quality of all buildings, especially workplaces, and healthcare and education buildings. Feedback from the building users and experts should feedforward to future industry-wide projects.

As I commented in an article in 2018: "it appears that POE is still very much on the agenda of many organisations in the UK. The conclusion is, therefore, that POEs are being carried out, but just for 'personal consumption' and not shared among the workplace community". More active sharing of POE findings with peers, making it routine and expected, will encourage more uptake and help lessen the above barriers and concerns.

The reported POEs need to be candid, sharing the lessons learned and details of the areas for improvement rather than simply being a showcase case study. Lessons learned from buildings that have been found to perform poorly, rather than just those that perform well, are particularly beneficial. The key is for occupiers, designers and facilities managers to gather feedback on the building's performance and share it with their peer group so that building design and operation can improve, thus providing better buildings for business in the future.

I appreciate that conducting and sharing regular POEs is a "big ask", but it is the responsibility of all real estate professionals to endeavour to ensure the buildings they design, construct or manage meet the needs of the occupants, and POE is an essential and practical means to achieve this.

Hopefully, this book will encourage the take-up of POEs and, by using sets of common questions highlighted later, it should be easier to compare buildings within a portfolio and across numerous clients and building types. Indeed, future POEs should be reported as short case studies, similar to those in Chapter 10, providing an overview of the performance of the building and identifying any lessons learned.

Note

1 https://well.support/pre-approved-programs~ca36f076-229e-438f-b23f-643626026f74.

3   When to conduct a POE

The next question after "What" is a POE and "Why" conduct it is "When" should a POE be carried out. The answer partly depends on the type and level of POE, but as with all things POE, there are contrasting but overlapping views on when to conduct one, as outlined below.

There is merit in conducting a pre-project POE to establish a baseline and inform the design brief, but the trickier question is when to conduct a POE after project completion.

Timeframe for POE

The now superseded Office of Government Commerce (OGC, 2007) suggested that building performance measurement is sub-divided into the three following stages.

1. **Project evaluation** – A quick check of how well a project is performing during the design and implementation phases.
2. **Post-project review** – An assessment conducted soon after project completion, which focuses on how well a project was managed, such as completion on time and on budget, and the project team's performance.
3. **Post-implementation review** – An evaluation carried out some time after a project has been implemented, used to determine whether the project has benefitted the occupying organisation.

The OGC's third stage, post-implementation review, is the one most in line with a POE, but the earlier evaluations are also worth including when assessing project performance. Indeed, some project sponsors are keen to see early results. Likewise, the IPA (2021) proposed a review when the project is about to handover to the organisation for normal operation and then again 6–12 months after project completion. The IPA also recommended that reviews are carried out throughout the life of the building.

HEFCE (2006) also proposed three stages of the review process for higher education institutions. Its project review, like the post-implementation review, is what

DOI: 10.1201/9781003350798-3

is usually considered to be the timing of a typical POE. HEFCE's stages, below, focus on collating occupant feedback but can involve technical POE techniques.

1. **Operational review** – Carried out 3–6 months after occupation. Once the occupants have settled in and familiarised themselves with the building, they can be asked about how well it is working and whether there are any immediate problems that need resolving.
2. **Project review** – Conducted 12–18 months after occupation. Usually carried out after at least one year of occupation when the building's systems have settled down and there has been a full seasonal cycle. This gives the opportunity to see how the building performs under a variety of conditions. It also gives building users a chance to identify where the building does not meet their long-term needs.
3. **Strategic review** – Carried out 3–5 years after occupation. This would take place several years after initial occupation when the organisational needs may have changed, and the building does not necessarily meet them anymore.

In my opinion, collating occupant feedback within three months of project completion is too soon because the building may still be undergoing commissioning or snagging, affecting the occupants' views. Alternatively, the staff may still be wowed by their new environment, referred to as the "honeymoon period". A light touch survey shortly after occupancy capturing views on the smoothness of the move/transition and initial impressions may appease eager project sponsors, but it is usually better to wait.

In its report, *Government Soft Landings*, the UK BIM Framework (2019) advised that "the timing and format of the POE should be established as part of the GSL strategy and implementation plan". It also referred to *BS 8536,* which states that the formal POE of a building's performance should be conducted at the end of years one, two and three. *BS 8536* distinguishes between initial and extended aftercare: "The ongoing support of design and construction delivery team(s) during the extended period of aftercare (typically three years) enables measurement of performance targets and fine-tuning of systems".

RIBA (2020b) seems to follow GSL guidance but suggested that a light touch "simple but meaningful rapid evaluation" is undertaken post-occupancy and before the building contract concludes, warning that: "This level may not fully reflect the true building performance as it may not include full seasonal information. However, it can provide some initial useful insights for the client and offer some feedback for future projects". Furthermore, the RIBA report stated that a more detailed diagnostic "POE usually occurs in the second year of occupation and verifies the performance of a building and reviews any issues discovered"; and finally, that a "detailed (forensic) POE can occur at any time, but ideally [should] be completed by the end of the third year of occupation".

The Skills Funding Agency (2014) mandated that all organisations to which it provided capital funding are "required to provide an evaluation, to include functional performance, between 12 and 18 months from occupation, when the impact that the project has on educational delivery, finances and estates performance is known". Similarly, a point is awarded in the BREEAM certification to those who show a commitment to carry out a POE one year after the initial building occupation. The Queensland Department of Housing and Public Works (2017) recommended that a POE be conducted for a new or existing facility when it is fully operational, typically after at least 12 months of occupancy.

Preiser and Schramm (2005) proposed that "POEs are ideally carried out at regular intervals, that is, in two- to five-year cycles, especially in organizations with repetitive building programmes, such as school districts and federal government agencies". There is certainly merit in conducting regular, say annual, POEs of occupied buildings to understand how the space is responding to the changing needs of the occupying organisation over time and to support facilities managers in proactively operating and maintaining their buildings. The information from repeat longitudinal POEs can be captured to create a database that allows a comparison and benchmarking over time.

As well as the frequency, the time of year that a POE takes place also needs consideration. It is useful to know how the building performs across all seasons, especially the more extreme summer and winter ones, which implies leaving conducting a post-project POE for up to one year. However, I prefer to launch one-off occupant feedback surveys in spring or autumn so they are not heavily biased by any recent extreme conditions. Having said that, the purpose of the POE may be to better understand how the building performs in extreme conditions, in which case the timing should be appropriate. Furthermore, a POE should be timed to avoid periods of low occupancy, such as holiday periods, conferences and team away days. Of course, with the increasing use of sensors, some components of a POE, such as indoor environmental monitoring and occupancy, can be measured continuously in real-time.

So, there are some differences in opinion of when to conduct a post-project POE. Based on the breadth of literature and my experience, my advice is as follows and illustrated in Figure 3.1.

- **Operational review** – Conduct an operational review shortly after project completion. This is a light touch assessment involving the project sponsor and team on how well the project was managed, using metrics such as cost and time, how the project team performed together, and so on. It may involve a walkthrough, workshop, review of relevant documentation and views of key stakeholders.

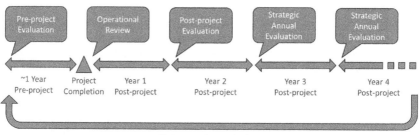

Figure 3.1 POE throughout the building life cycle

- **Post-project evaluation** – Perform a post-project evaluation, that focuses on feedback, after occupants have spent a summer and winter in the new or refurbished building. This is usually conducted around 12 months after project completion. This has the advantage that the occupants have experienced the more extreme seasons, but any initial teething or commissioning problems have been resolved. Project sponsors and designers usually expect feedback on their project within one year while the project is still topical. Any problems flagged by the occupants may need more detailed follow-up technical reviews. POEs focusing on actual compared with predicted sustainability performance usually require 12 months of data but, if necessary, occupant feedback may be collated slightly earlier, after around 9 months.
- **Strategic annual evaluation** – Carry out regular annual POEs starting two years after initial occupancy. Particularly with a major refurbishment or relocation, it is advisable to conduct a strategic review once the building design and operation is considered "business as usual". Projects that are required to conform to GSL and/or BS 8536 are also required to conduct POEs in years two and three post-occupancy. These later evaluations may be more technical and detailed, depending on the POE objectives and results from earlier studies.

Planning a POE

The timing of a POE is also dependent on the planning of it. Below are some points, identified by Kennet (2021) and myself, to consider when planning a POE.

- Determine the internal sponsor who will take ownership of and drive the POE.
- Define the purpose and objectives of the POE, clarify the purpose and what is hoping to be achieved. Identify the building elements and systems to be evaluated.
- Agree on the POE metrics and data-gathering techniques that are required to achieve the objectives. Use an established methodology if you want to be able to benchmark or are short of time. Alternatively, develop and pilot the methods to be used, especially occupant feedback surveys.
- Confirm the timing of the POE and the different components. Consider if it is a one-off evaluation or continuous monitoring.
- Select the internal person or team capable of conducting the POE or commission external advisers specialising in POE or elements of it, such as feedback surveys or indoor environmental monitoring.
- Select the occupants, other building users and key stakeholders to be involved (see Chapter 4).
- Communicate clearly when the POE is taking place and the reasons why. Encourage the occupants to offer feedback.
- Consider how the results will be collated and shared with the project team, occupants, other stakeholders and the wider building industry. Confirm any report deadlines and arrange any relevant events.

4 Who to survey

You may recall from Chapter 1 that Preiser, Rabinowitz and White's (1988) original definition of POE was "the process of evaluating buildings in a systematic and rigorous manner after they have been built and occupied for some time" and Preiser and Vischer (2005) later proposed that POE "addresses the needs, activities, and goals of the people and organizations". Collating the feedback from a representative sample of the building's users is clearly a fundamental element of POE.

Building users

The building users comprise the following groups.

- **Occupants** – Clearly, collecting feedback from the primary occupants is key. However, it is important to consider the roles and activities of different occupants; for example, the views of admin support staff often differ from those of senior management. As well as role and seniority, there may be important differences between the views of people of different gender, age, culture and tenure etc. The facilities team and others responsible for the workplace, such as the HR and IT departments, will have a different perspective from their colleagues. Furthermore, as the roles and activities of different teams and departments vary then so may their response to the building. Note the terms "occupant" and "occupier" are often intermingled, but I tend to consider occupier to refer to the organisation and occupant to the individual.
- **Key stakeholders** – This common term is often used in POEs, where it refers to the people who are highly interested and invested in the success of the building project. This usually includes the project sponsor, project design team and leadership team, but in the broader sense may be extended to employees, customers, shareholders, suppliers and communities. Which key stakeholders need to be consulted will depend on the type of building and magnitude of the project. The key stakeholders closely involved in the project are likely to have different, possibly biased, views compared with their colleagues, but they will also be more informed of the project objectives and details. The feedback from different stakeholders is sometimes weighted, by relevance and importance, to help achieve a more representative view of the project.

DOI: 10.1201/9781003350798-4

- **Visitors** – Most buildings receive some visitors and guests. Depending on the building project, the views of visitors may also be considered important. For example, corporates will want their visitors, particularly clients, to feel welcome and impressed. Those in public sector buildings, such as municipal buildings and public libraries, will also want their spaces to meet the needs of their visitors.
- **Local community** – The general public may provide feedback on the buildings they regularly visit, but they may also have views on other buildings in their local area. The views of the local community and how they are affected may be important for some projects, usually those that are high-profile and large-scale.

Deciding who to gather feedback from in order to generate a representative POE is crucial because the building users vary with each organisation and space. For example, in higher education institutions feedback from the students is also sought as well as from the academic and professional services staff. My own POEs show that these groups usually have very different opinions on the success of a building project. The views of students' parents may also be relevant, especially if they are involved in the process of choosing a university. Universities will also receive a variety of visitors, from the general public at events to local, and possibly international, dignitaries.

Prisons and police stations are also places with a range of building users who are likely to have very different views. If evaluating prisons, consider the views of the guards, the support staff and the prisoners; and for police stations seek feedback from the police officers, civilian support staff, visiting public and maybe those detained for questioning or held in custody.

Survey respondents

Occupant views will vary based on their role, activity, project involvement and sociodemographic factors, among other things, so it is essential that the people participating in the POE feedback surveys are representative of all building occupants.

Sampling and target population

There are different means of seeking occupant feedback, including interviews and focus groups, but it is questionnaires that tend to have the widest reach. It is, therefore, particularly important that the questionnaire respondents are representative of the "target population".[1]

It is usually easiest, both logistically and politically, to include all occupants in a feedback survey. However, if this is impractical, perhaps due to budget or time limitations or because there are a lot of scattered occupants, then a sample[2] of the population could be invited to participate in the survey. However, it is important to ensure that the sample accurately represents the target population, otherwise the sample may be biased in some way and a biased sample will produce biased results!

Social scientists use three main techniques when selecting a survey sample.

- **Random sampling** – Participants are chosen randomly so that they all have an equal and known chance of being selected. For example, randomly selecting 200 occupants from a staff list of 2,000 gives each occupant an equal 1 in 10 chance of being selected. If the survey is issued for a limited time, then some occupants may be absent and not have a chance to respond so, strictly speaking, this would not be considered random sampling.
- **Systematic random sampling** – This involves choosing every N^{th} person from a random list after starting in a random place, where N is the total number in the target population divided by the required sample size.
- **Stratified random sampling** – Here, the target population is divided into sub-sets, or strata, and then random sampling is applied to each of the sub-sets. For example, if it is important to ensure that all job roles are represented, then the sub-sets might include senior managers and administrative staff or perhaps represent the different departments. The number, or quota, selected from each sub-set is then made proportional to the actual size of each group. This can become complex when stratifying across several sub-sets.

When discussing the sample, it is important to check that the sample represents the different locations in the building, for example, each floor and the desk positions on each floor. If appropriate, the sample should also represent the various business units or departments, grades or job functions, and desk types (for example, open plan or private office). This may mean that it is necessary to set a quota and use stratified random sampling to ensure sufficient respondents are selected in each category. The simpler alternative is to invite all building occupants to participate in the feedback survey.

On a related note, sampling also applies to indoor environmental measurements. It is usually impractical, and too costly, to monitor all the building spaces. The location of the monitoring equipment will need to be agreed upon in advance of the POE, as will the length of time measuring. Often, the equipment is moved around to different parts of the building to make one-off measurements or is left for a short period of monitoring.

Sample size

Another frequent question asked when conducting occupant feedback surveys is, "How many people should be included in the sample?" This depends on the target population and the level of precision required, which social scientists express using the "confidence level" and "confidence interval". The confidence *level* is a measure of certainty, that is how closely the survey sample reflects the answers that the whole target population would have given. It is typically quoted as a 95% confidence level, or in other words, 95% certainty that the population would have chosen the same answers as the sample, equating to a 5% chance. The confidence *interval* is the percentage margin of error in the confidence level. For example, if the reported confidence interval is 5% and half (50%) of the sample chooses a particular answer then the chance that the whole population would have chosen the same answer is between 45 and 55%.

Figure 4.1 illustrates the sample size required for a particular population to provide the conditions usually aimed for by social scientists: a 5% confidence interval and a 95% confidence level. For example, a target population of 200 occupants would need a sample of 132, whereas a target population of 2,000 occupants requires a sample of only 322, and above a 2,000 population the sample size flattens. This is counter-intuitive for most people, who tend to assume that much larger sample sizes are required for the research to be statistically robust. Consequently, the results from small sample sizes may be dismissed as inaccurate. However, most statisticians agree that the minimum sample size to achieve a meaningful result is around 100, with 200 being more acceptable. However, the response rate (discussed next) may mean that more occupants need to be invited to participate to be sure of meeting the required sample size.

The level of breakdown of the data, or the analysis of sub-sets, also has an impact on the sample size. For example, when comparing the views of the younger and older generation, the survey sample may be split into those aged 30 years or less, then those 31 to 40 years of age, 41 to 50 years old and those aged over 50. To ensure statistical robustness there needs to be 30 or so people in each group, sometimes called a cell. So, the total sample size would be 120, but bear in mind that more people may need to be surveyed to achieve this sample size. If a two-way breakdown is required, say age by role, then the number of cells will be multiplied.

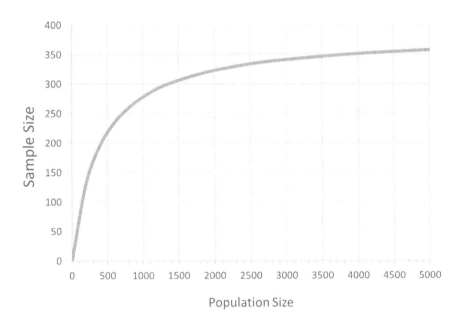

Figure 4.1 Required sample size based on population size

For example, the four age categories and, say, five job roles result in 20 sub-groups/ cells and a sample size of 600. If there are insufficient responses in each cell, then they may be combined to create larger sub-groups for analysis, for example, 40 years of age or less and above 40 years old.

The key to reducing sample bias is to create a representative sample. That depends on the sample size, the sampling procedure described earlier and the response rate.

Response rate

Not only is it important to approach a representative sample of occupants, but it is also important that those approached then respond and complete the feedback survey. A good response rate (RR) is required, where the RR is the ratio of the questionnaires returned to those invited to participate, expressed as a percentage:

$$RR = \text{Returned} \div \text{Invites} \times 100 \text{ (\%)}$$

Note that partially completed questionnaires are usually regarded as a response, although partial completions do mean that less useful data is collected and available for analysis.

A low RR can lead to a biased sample – people who are keener to respond may have a vested interest in doing so. In theory, people with complaints about the building are more likely to respond, so the survey results may make the building's performance appear poorer than it is perceived to be by a representative sample of the building population. However, in practice, it is those with something to contribute, whether positive or negative, or those most affected who are likely to respond. Also, too few responses may result in there being insufficient data to conduct meaningful analysis, particularly if specific sub-groups are being compared. Nevertheless, remember that a small representative sample is more acceptable than a large nonrepresentative one.

The next step is to consider what is an acceptable response rate. Unfortunately, there is no simple answer because it depends on the type of survey and how accurate the responses need to be. The more I research response rates, the bigger the range of figures I find. For example, Bunn (2020) suggested that: "Response rates below 40% may be evidence of patchy spatial coverage. Response rates of 30% and below may indicate serious shortcomings with awareness-raising and/or survey technique".

My personal view of response rates, derived from a collation of different researcher views, is as follows.

- When the RR is less than 33% (one-third), there is a very high chance of sample (non-response) bias, and this may result in insufficient numbers for comprehensive analysis. Extra effort should be made to increase the RR or the survey results should be disregarded or, if necessary, presented with a caveat of low confidence.
- A RR of 33–50% (one-third to one-half) also runs the risk of having a biased sample. The survey results should be presented with a caveat of low confidence or, preferably, more responses should be gained.

- A RR of 51–74% (over one-half) reduces the risk of a biased sample and the results can be presented with some confidence. While 51% is technically a majority, nevertheless, it is definitely worthwhile taking additional time to increase the RR.
- 75% (three-quarters) or higher is a good RR, where the sample is unlikely to be biased and the results can be presented with confidence. 75% is a practical and achievable response rate for surveys of occupants in office buildings.

If the building users do not return a questionnaire, it does not mean they have rejected it: they may require more time to respond. A low response rate may be due to several, somewhat different, factors that can be readily overcome.

- **Bad timing** – When inviting staff to participate in a survey, try to avoid holidays and other days when staff are less likely to be in the office, such as away days, Fridays or when there is high absenteeism due to sickness. To help increase the RR, keep the survey open for two or possibly several weeks with regular reminders.
- **Survey fatigue** – This is partly related to timing. I started this book by highlighting that feedback is ubiquitous. Some organisations may conduct regular surveys, so try to avoid clashes in timing. Also, keep the survey short and to the point.
- **Too long** – The most common reason for people not participating in, or not completing, a survey is that it is too long or too complicated. The survey needs to be relevant and to the point.
- **Inappropriate medium** – Feedback can be collected using different media. An emailed online survey will work for most, but also consider paper-based, kiosk-style, postcard or app-based surveys, especially for building visitors.
- **Organisational culture** – Sadly, the response rate may be low due to organisational issues. The participants may feel that the survey is pointless and a tick-box exercise where their views will not be listened to. Others may be concerned over the consequences of answering a survey truthfully but negatively. The latter can be helped by communicating that the surveys are anonymous with only grouped data presented. In contrast, I have achieved 98% response rates in organisations with a positive corporate culture.
- **Poor communications** – Prior notification and the invitation to participate in the questionnaire survey are crucial for increasing the response rate and deserve the more detailed explanation given below.

A well-constructed encouraging invite, sent from the leadership team and/or a respected colleague, will increase the response rate. The invite should:

- briefly describe why the survey is being undertaken,
- explain why the participant was chosen and justify why the participant should complete the questionnaire,
- explain how the results will be used and the next steps,
- clearly state the survey timescale and deadline for completion,

- explain that the survey is being conducted with prior management approval, so the participant can spend (paid) time responding to the questionnaire,
- assure the participants of their anonymity and guarantee confidentiality, explaining that no individual responses will be passed on to management and only grouped averaged and anonymous data will be reported,
- provide a clear link to the online survey or the instructions for other feedback mechanisms,
- provide a contact name for any queries and name the project sponsor.

If the response rate is low, then those who have not responded will need chasing. Typically, two reminders are sent out, one just before the survey deadline and another just after. If the response rate is still low, then the survey period might be extended with more reminders. A personal follow-up or phone call is better than an email for encouraging participation, although this is not always practical. If the survey is anonymous, then the reminder will need to go to all participants rather than those recorded as not responding. An alternative is to ask the business group head or departmental administrative support to chase responses.

There is some debate among researchers over whether incentives should be offered to motivate a response, for example, a gift voucher or prize draw. In POEs the participants usually have a vested interest in responding, so offering to publicise an overview of the results and use the results to determine any valid ideas for following up is usually sufficient to improve the response rate. Offering incentives such as payment is likely to result in a disinterested or biased sample. However, offering a free cup of coffee or similar upon completion of the questionnaire is a nice "thank you" gesture.

Who conducts the POE?

As well as who to survey, I need to mention who is best suited to conduct the POE. The intention of this book is to show that it is possible for a POE to be conducted by an in-house colleague, especially if the POE is at a high level and mostly involves occupant feedback. However, it is advisable to use people with expertise in questionnaire design and in conducting interviews to facilitate focus groups. Obviously, if more technical techniques are to be implemented, then it is usually necessary to employ people with the relevant skills and experience, as well as the equipment.

It is also a good idea to commission an independent adviser who has less vested interest in the project and whose views are not biased. This is not always practical: for example, some design teams may have their own POE advisers and offer their services as part of their design programme even though they are not disinterested parties.

Notes

1 The target population is all those invited to participate in the survey. It is the group of people whose views are being canvassed, usually the building occupants rather than visitors.
2 A sample is a sub-set of the target population. The survey sample is everyone who responds to the invitation to participate by completing and returning the questionnaire.

5   How to conduct a POE – stakeholder feedback

This chapter focuses on methods for collating feedback, mostly from the building occupants but also from visitors, workplace experts and the project team. Most POEs include subjective occupant feedback that can be obtained through interviews, focus groups and questionnaires, with each technique having its own advantages and disadvantages. However, an expert walkthrough is also highly recommended.

Interviews

Interviews are the process of obtaining feedback by talking directly to a building user. Interviews enable qualitative comments to be captured that can be used to verify and personify more quantitative survey results. Interviews are particularly good for covering sensitive topics or probing for in-depth information and detail. Interviews can include specific questions but work better with more open-ended questions; they can be used for exploration and capturing unanticipated comments and insight.

Interviews are the preferred methodology for investigating the organisation's objectives and vision with the leadership team (executive board members and/or senior managers) and for obtaining their feedback on issues such as value for money, whether the workplace represents the corporate brand, the impact on visitors, and the morale of staff.

In my experience, interviews elicit a higher response rate from the leadership team than questionnaires, and one-to-one interviews are usually easier to arrange than a focus group that depends on coordinating several diaries. Interviews are also useful for discussing operational and technical details with the facilities, HR and IT managers.

There are some disadvantages to interviews. If they only involve the leadership team, then a limited view of the overall performance of the building will be gained because they may not use the building so often or their activities differ from most staff. Likewise, interviews with project team members only will result in a biased view of the building's performance. Unlike questionnaires, interviews do not offer anonymity, but of course, the comments can be kept anonymous.

DOI: 10.1201/9781003350798-5

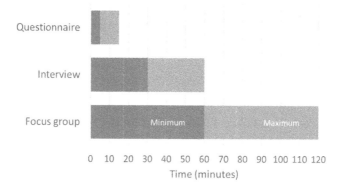

Figure 5.1 Participation time in the main feedback methods

Conducting interviews is time-consuming (see Figure 5.1), with each one typically taking 30 to 60 minutes to complete plus write-up and preparation time. However, it should be noted that designing a questionnaire and a focus group session is also time-consuming.

Interviews require some preparation and there is a skill, learned over time, to conducting them well. They are generally classed as "structured" or "unstructured".

- **Structured interviews** – these are most useful when the same information is required from each respondent and, therefore, the same questions are repeated in each interview. Sometimes a full questionnaire might be administered in person, but usually an interview schedule (crib sheet) is created as an aide memoire to the interviewer.
- **Unstructured interviews** – these may be conducted when exploring a new topic or when different information is required from each respondent. An unstructured interview may be useful to inform the design of a questionnaire.

Conducting a successful interview requires some basic skills and experience, but some basic techniques can be quickly learned.

- **Preparation** – Decide whether to send the questions to the interviewee in advance. Personally, I send out an agenda or broad themes rather than specific questions. However, if specific information is required, such as headcount data, then ask for this in advance.
- **Introductions** – Before the interview, check who is being interviewed and that their roles and interests relate to the project. Start the interview with introductions followed by the interview objectives.
- **Ground rules** – Explain the "rules of engagement", such as whether the interview is "off the record" or will be transcribed and reported. Explain whether the responses are to be presented verbatim and ascribed to the interviewees or paraphrased and anonymous. Let the interviewee know if the interview is being

recorded, if required for transcription or checking comments. Some speech-recognition software and virtual meeting platforms allow recordings to be automatically transcribed, saving time in capturing interviewees' verbatim comments.

- **Ease in** – Put the interviewee at ease: start with an easy question. Perhaps ask the interviewee to describe their role and that of their team, or maybe ask them about their day at work or what is currently occupying their time. Consider a relevant icebreaker question.
- **Open-ended questions** – Interview questions are usually quite broad and open-ended, requiring longer responses. However, sometimes it is worthwhile asking a question that is to be answered using a showcard with a rating scale on it, like those used in a questionnaire survey. This is a means of quantifying some interviewee responses. The questions might be the same as those used in a questionnaire issued to all occupants so that the interviewee responses can be compared, especially if they are members of the leadership team who did not reply to the questionnaire survey.
- **Explore** – If the interview is structured, then keep to the script but also allow time for the interviewee to expand on their responses and be prepared to explore and deviate a little before returning to the script. Coax the interviewee to answer the core questions.
- **Listen and confirm** – Allow the interviewee time to speak and reach their point. A good interviewer listens more than speaks. Take notes to show that you (the interviewer) are listening and repeat back or summarise key points. Do not be afraid to ask for responses to be repeated or clarified, and challenge any responses that may be contradictory.

In my opinion, interviews are better conducted face-to-face but, when there is limited time or difficulty arranging in-person appointments, an online (using Team, Zoom or similar platforms), or even telephone, interview may be conducted.

Interviews are often one-to-one, but having two interviewers often works better – taking it in turn to ask and record questions. Similarly, there may be two or even three interviewees, but the larger the group, the less likely it is that personal opinions will be openly shared. More than three interviewees is akin to a focus group or workshop.

Focus groups and workshops

Focus groups and workshops both bring together a group of stakeholders for an in-depth discussion of topics, but there is a subtle difference between the two. Focus groups tend to have a moderator that steers a discussion around a particular topic eliciting answers to specific questions, whereas workshops tend to be more exploratory in nature with collective thoughts and opinions (Figure 5.2). Regarding POE, the difference is immaterial but a post-project gathering is more like a focus group discussing the success and failures of a project. In contrast, a pre-project gathering is likely to be more exploratory, informing the design brief so it is more like a workshop, although problems and solutions could be explored post-project. In this section, I will focus on post-project evaluation and therefore use the phrase "focus group".

Figure 5.2 Author leading a workshop on designing for psychological needs

The advantage of focus groups over one-to-one interviews is that the views of several stakeholders can be obtained in a similar amount of time. The participants will interact and build on each other's comments, possibly forming a consensus. Focus groups are often combined with a questionnaire to gather more detailed information. For example, questionnaires are useful for quantifying (or scoring) the views of many occupants, but the score does not necessarily divulge the underlying reason for that score. A focus group conducted after the questionnaire could discuss the survey results, which should help to explain any key issues and assist in producing recommendations. A focus group is also quite flexible, and any issues missed in the questionnaire could be discussed.

There are also disadvantages to focus groups. Notably, they take much longer than completing an online survey, despite the participant needing little time to prepare. The participants need to be carefully selected to ensure that opinions are not biased. A series of focus groups with a range of building users may be required, for example, with people from different roles or departments. Furthermore, the opinion of one or a few participants can overwhelm that of the others present and, therefore, may appear to be a majority opinion.

The person or team running the focus group(s) will need to be well-briefed with adequate facilitating skills and experience. The output from the focus group is mostly qualitative information, and it is more laborious to systematically analyse focus group data than data collected through a questionnaire. Furthermore, some clients and researchers may consider such qualitative data to lack statistical rigour.

Focus groups work best with 8–12 participants, coincidentally the size of a "squad", the smallest military unit. This size allows time for each participant to introduce themselves, get to know the others and be actively involved; it is also easier for the facilitator to manage and involve lower numbers of participants. The attendees may be a representative mix of occupants or a particular target group, such as senior managers or administrative support. However, limitations on the availability of some people can lead to an unrepresentative group of participants at the meeting.

I prefer to conduct focus groups in person rather than online. I appreciate this is not always practical because it takes longer to set up and to conduct (especially if travelling), but my experience is that the quality, engagement and outcomes are better.

I often refer to the quote, "Tell me and I forget; teach me and I may remember; involve me and I will understand", ascribed to Confucian philosopher Xunzi and sometimes to Benjamin Franklin. Focus groups and workshops benefit from interaction and engagement, and this is more successful and easier if conducted in person using a range of hands-on techniques and materials. Nevertheless, there are some platforms for supporting online group sessions, such as Miro[1], Mural[2] and Microsoft Whiteboard.

There is already much guidance on the techniques and exercises that can be used in group meetings to provoke useful responses. Pre-project workshops probably require more creative techniques to elicit exploration and a broader range of ideas than a post-project focus group. Nevertheless, it is useful to use a range of techniques when facilitating such meetings and there are some generic techniques useful in all situations.

- **Introductions** – Ask each participant to introduce themselves, say why they are at the meeting and, possibly, what they hope to get out of the meeting.
- **Icebreaker** – To put the participants at ease, it is always good to start with an icebreaker. This is usually a short and fun exercise that may be loosely relevant such as "tell the group about your favourite place" or more tangential such as "If you were stranded on a desert island what would you want with you?"
- **Call-out** – For some questions, and if short of time, it is simplest to ask participants to all out answers. Just be careful that one or two participants are not the only ones answering. If so, then select and invite others to comment.
- **Round robin** – If there is more time, ask the participants in turn to respond to a question. This ensures that all are involved, and the responses are likely to be more representative.
- **Brainwriting** – Brainstorming involves calling out and sharing ideas. Brainwriting is when ideas are written down and shared, usually on sticky notes. This exercise allows the participants to have a little more uninterrupted time to think. It is popular with people who are less vocal or who wish their comments to remain anonymous. The notes are displayed and talked through, allowing more related/prompted comments to be added.
- **Images** – Using photos as prompts can be useful in focus groups. A range of images of the project could be collated, and the participants asked to comment on them. Alternatively, with planning, the participants could be asked to take photos of the elements of the project they like and dislike or, pre-project, bring along photos of other projects and design ideas they like. Photos are predominantly used to prompt discussion but they may also be analysed for common themes, currently manually but possibly by AI in the future.
- **Breakout groups** – For longer sessions splitting the participants up into smaller discussion grouse helps break up the format. It is also a useful technique if a number of topics are to be discussed with limited time, so each group can discuss a different question. Remember to ask the groups to report back on their discussion.

As mentioned, running a focus group requires an experienced facilitator. A good facilitator will always:

- plan ahead and prepare the meeting materials,
- get to know the participants and build their trust,
- set the scene, share the agenda and define the workshop purpose,
- explain the ground rules (the "rules of engagement"),
- facilitate the meeting, keeping it on track,
- encourage all participants to be involved, and the more vocal of them to pause for others,
- use a range of workshop techniques,
- thank the participants and communicate the next steps.

Questionnaires

I believe that occupant feedback is the most fundamental element of a POE, and you will notice that the main thrust of this book suggests using questionnaires to gather feedback. More specifically, the focus is on the use of self-completion questionnaires, rather than those administered via interviewers, which are more commonly used in public surveys, for example, when canvassed at home by a field operative for opinions on broader issues.

Pros and cons of questionnaires

There are many advantages of self-completion questionnaires over other feedback techniques, including:

- they are a cost-effective means of gathering information from a large group of people using multiple-choice responses (rating scales) to provide a broader-based opinion,
- they are the least intrusive way of gathering feedback because people may choose whether or not to participate and can usually complete the survey at a time of their own choosing (within limits),
- they are usually less time-consuming for the individual respondents compared with participating in an interview of a focus group,
- respondents are usually asked to choose from a selection or range of answers, so it is relatively simple to compare (and benchmark) everyone's quantitative replies, whereas other techniques that are gathering opinions, such as unstructured interviews, may produce a wide and incomparable set of responses,
- due to their low cost, self-completion questionnaires can be administered to all participants, the target population, which is more likely to yield a larger sample than other methods which target a select group of stakeholders,
- participants can remain anonymous; however, if sufficient background questions are asked it is possible to pinpoint the issues raised on specific areas in the building.

Building on the last point, the identity of the respondents can be protected in all feedback methods, even interviews. However, it may be more useful to know the identity of the respondents so that any particularly interesting or unusual responses can be followed up, or to help solve any problems that are identified. As a compromise, providing names is usually optional. However, questionnaire responses should always be stored without names to conform to local data protection laws.

As with all methods, there are some disadvantages to self-completion questionnaires:

- it is possible for a questionnaire to only elicit responses from particular groups within the target population; for example, senior managers may not have the time or inclination to respond, so the representativeness of the survey sample should be checked,
- questionnaires that are administered personally by an interviewer or by phone survey can produce a more representative sample because specific occupants can be repeatedly targeted and then assisted with completing the questionnaire once they agree to be interviewed,
- although self-completion questionnaires will provide quantified data, such as the level of dissatisfaction within a group, this does not explain the cause of the dissatisfaction, so ideally self-completion questionnaires are followed up by focus groups where the responses to the questions are probed for detail,
- skills are required to design, administer, analyse, interpret and report on questionnaires.

Length of questionnaire

A higher RR is key to obtaining an unbiased sample. If only a small number of completed questionnaires are returned (such that the RR is low), there is a danger that the results will not represent the full population or are based on a biased sample.

A well-designed and implemented questionnaire is essential for producing a high RR and avoiding sample bias. The main factor that affects the RR is the length of the questionnaire. There is no clear guidance on the length of questionnaires, but experience shows that for workplace evaluations a self-completion questionnaire that takes approximately 10–15 minutes is acceptable.

Questionnaires provide subjective information – that is, they report opinions and estimates rather than hard data – so it is important to reduce the margin for error by developing well-structured and well-written questions that are reliable and valid. Reliability is when a question can be asked repeatedly and it produces consistent results, whereas validity is when the response to a question reflects what it is expected to measure. For the responses to be valid, the questionnaire must be reliable, but reliability does not guarantee validity! The wording of the questions is crucial to ensure reliability and validity, and that will also help increase the RR.

Online versus paper-based surveys

A self-completion questionnaire may be administered online, through email, the internet or an app, or it could be paper-based and delivered via a desk-drop, pigeon-hole or even by post. Figure 5.3 shows the same questionnaire in its original paper and later online format. When online questionnaires were first introduced, the response rates were quite low, but now the RR for online surveys is comparable and often better than that of paper-based surveys. However, a few organisations still prefer paper-based surveys.

The main advantage of a hand-delivered paper-based questionnaire is that the survey administrator can briefly meet the participants, allowing them to have a quick chat about the building and pick up qualitative information. They can also build a rapport and encourage the participants to complete the questionnaire, thus helping to increase the response rate. Historically, paper-based surveys resulted in higher RRs than online ones, but that is no longer the case: I have achieved a 98% RR with an online survey, and the techniques highlighted in Chapter 4 are more pertinent to higher RRs. Although less common now, some companies still do not offer internet access to all their staff, which would restrict the use of online questionnaires.

The main advantages of web-based surveys over paper-based questionnaires are that they:

- are a more cost-effective way to administer a questionnaire to all the occupants,
- have a fixed cost because the cost does not increase with an increase in sample size (unless there are many open-ended questions which take longer to analyse),
- are relatively low cost to implement because the data is automatically captured and formatted,
- are quicker to report back because there is no need to re-enter the data,
- are quicker to launch because printing, distribution and collection are not required,
- allow question routing (filtering and skipping) so that the respondents do not waste time answering irrelevant questions,
- offer a range of formats – most proprietary online survey development platforms offer rating scales, check boxes, rank ordering, slide bars and so on,
- allow the order of questions or answer options to be changed to prevent bias due to the order they are presented,
- allow respondents to pause part-way through a questionnaire and recommence where they left off, but this also applied to paper-based surveys,
- prevent people from responding multiple times, either from using copies of the same paper-based survey or completing an online survey on the same computer,
- can be administered in real-time, for example some researchers offer short questionnaires that pop-up on all the occupants' computers at specific times, which is useful when marrying occupant responses to recorded environmental conditions,
- can be administered in high-security environments where external survey staff are not allowed access.

Survey development process

I have tended to use the terms "questionnaire" and "survey" interchangeably, but technically *questionnaire* relates to the content (the questions) whereas *survey*

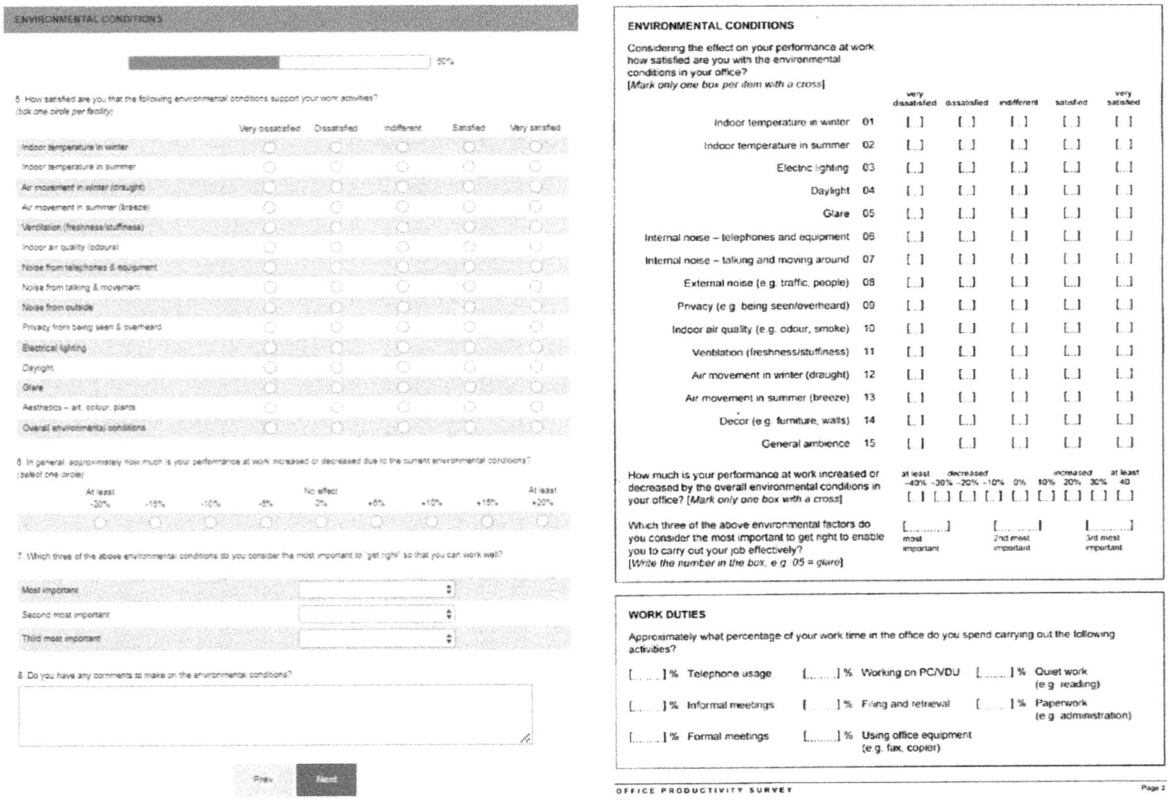

Figure 5.3 Online (left) and paper-based (right) versions of a similar questionnaire

refers to the whole process from developing a questionnaire to reporting the results. Occupant feedback survey development is a multi-stage process requiring the following key steps:

- defining the objectives of the survey, which is crucial for developing the content and approach,
- agreeing on the feedback methodology, for example, an online questionnaire or other,
- selecting the sample, such as all occupants or a selection of them,
- developing the survey content, which may be an iterative process using existing or new material,
- piloting the survey, which is important regardless of whether using existing or newly developed wording,
- administering the survey, usually online but may be on paper or in person,
- analysing the data, for example, using charts to disseminate the survey findings,
- interpreting the results, using a combination of the survey data, qualitative input and walkthrough and so on,
- reporting and presenting the results, ideally in an easily digestible format sent to the project sponsor and also to the survey respondents (although the latter may be just headlines).

Figure 5.4 illustrates the process for conducting a post-project feedback survey, showing that the survey is usually accompanied by a walkthrough of the building being studied and followed up with qualitative research, such as interviews or focus groups to discuss the questionnaire results, assist in interpretation and input to a final report. The walkthrough takes place early to inform the survey design, and some early interviews would also help the survey design.

While new questionnaires can be developed, there is value in using tried and tested survey techniques. POE consultants may favour a proprietary questionnaire or may suggest using or adapting questionnaires developed by other organisations.

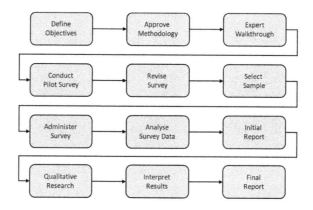

Figure 5.4 Feedback survey process

Expert walkthrough

The expert walkthrough basically involves the professional(s) conducting the POE, or possibly members of the design or facilities management team, walking through the building after project completion and making observations. The walkthrough may substantiate the feedback and output from the other POE methods or pick up on technical details missed by the occupants. A more thorough expert inspection may be undertaken of the building services, such as the PROBE studies.

A walkthrough with the project sponsor and members of the design team is a good approach because each discipline will notice different aspects of the building design. For example, I remember conducting a walkthrough of a building with my design colleagues on the day of practical completion. The lead designer was upset that the top of the screens between desks did not align with the storage units (there was just a 2 cm difference), but he had overlooked that some of the task chairs were missing.

During the walkthrough, the various elements of the building design could be captured using a short paper-based or online scoring/rating matrix (for example, see Figure 5.5). The quantified combined rating for each building design element can then be presented along with the qualitative professional comments. Photos are another useful way of documenting observations made during a walkthrough. As the adage goes, "a picture paints a thousand words" and a photo with a simple caption plainly verifies the subjective observations.

Workplace Quality Appraisal

Rating scale: 1 = Very poor, 2 = Poor, 3 = Satisfactory, 4 = Good, 5 = Excellent

Finishes	Rating	Comments
Ceiling		
Floor coverings		
Wall coverings		
Partitions		
Doors		
Furniture		
Finishes overall		

Furniture	Rating	Comments
Workstation surface		
Cable management		
Desk screens		
Chairs		
Desk layout/planning		
Meeting room layout		
Storage units		
Furniture overall		

Figure 5.5 Extract from an expert walkthrough pro-forma appraisal sheet

Some expert walkthroughs may involve a more detailed and technical assessment of the building components. For example, the PROBE evaluations that commenced in 1995 involved rigorous interrogation of the engineering infrastructure, in particular the building services.

Notes

1 https://miro.com/
2 www.mural.co/

6 How to develop a POE feedback questionnaire

This chapter expands on the previous one going into more detail on how to create a new questionnaire for obtaining feedback on how the occupants' current building supports them. Such a questionnaire may be administered after a move to a new building or post-project in a refurbished building. Alternatively, it may be used to evaluate the building occupied prior to a project; for example, to capture the requirements of a future building and inform its design. With careful design, the same questionnaire may be used pre- and post-project to allow a more accurate evaluation of the relative change as a result of the project.

Questionnaire completion

My experience indicates that for workplace evaluations, a self-completion questionnaire that takes approximately 10–15 minutes is acceptable. Too lengthy a questionnaire will result in a poor response rate and likely lead to a biased sample. Paying careful attention to the following points on questionnaire design will increase the likelihood that participants will complete the survey.

- **Number of questions** – Avoid asking too many questions by first agreeing on the objectives of the survey. The objectives help determine and constrain the questions required. Also, consider how the data from a specific question will be analysed, interpreted and used, and if unsure omit the question.
- **Question order** – The order of the questions can affect the response rate, and poorly ordered questions may result in partial survey completions.
- **Language** – If the questionnaire is to be applied internationally, then the language used may be a factor in the response rate. Questionnaires are more successful if written in the preferred language of those invited to participate. However, be aware that some words and phrases do not translate directly, so ideally the questionnaire should be checked by an expert in that language after translation or back-translated.
- **Format** – The presentation, writing style and ease of response are as important to the success of the questionnaire as the individual questions asked. Checklists of good survey design and well-written questions are provided below.

DOI: 10.1201/9781003350798-6

Too often, a one-off survey is seen as an opportunity for collating lots of information, resulting in an unwieldy questionnaire that does not adequately focus on the purpose of the POE. The objectives of the questionnaire should be discussed in advance: well-defined objectives usually lead to a well-designed and focused questionnaire. The objectives affect the number and type of questions asked, but poorly defined objectives also affect the analysis because it is difficult to interpret the results from a non-focused questionnaire. Simply put: if the survey objectives are unclear, then the results will also be unclear.

To some extent, the objectives will also dictate the order of the questions and the logical flow of the questionnaire. Poorly defined objectives may mean that key questions are overlooked. The questions asked should reflect the purpose of the study, and if the relevance of a particular question is unclear, then it should be omitted.

Feedback questionnaire content

Occupant feedback questionnaires can address a range of factors related to building use and performance. Such questionnaires tend to enquire about various building features and their impact on the occupants. Cohen et al. (2001) identified 12 topics on which questions would normally be asked in workplaces, including physical conditions within the environment, personal control over the physical conditions, management response to complaints, health and overall comfort productivity, background and the overall quality of the building. A more comprehensive but nevertheless non-exhaustive list is shown in the table. Most of the topics shown relate to offices, but some also apply to other working environments, including homes. More specialist buildings, such as hospitals, prisons, exhibition, teaching, research and performance spaces will have their own additional key building features for evaluation.

Clearly, the topics listed in the table are too many to include in a single feedback survey, and the list will need to be trimmed. The questions need to be tailored to best answer the POE objectives, address the building's main function and features and align with the organisation's circumstances. The POE questionnaire should also reflect the needs of the key stakeholders, for example a developer will be more interested in feedback on the base-build.

As well as deciding what features need to be evaluated, it is essential to have a clear view of how the respondents will assess them. Typically, the survey will address the occupants' experience and determine whether they are satisfied with the building features, or the occupants may rate how those features affect their comfort, wellbeing and performance. In addition to questions on specific building features, there are usually questions on overall satisfaction and performance and so on. Sections to allow free-text comments should be included – but not too many because they can take longer to analyse.

Sometimes occupiers request that certain questions are omitted from a standard tried and tested questionnaire because they consider them irrelevant. A common example is to exclude questions on catering if there is no restaurant provided. However, it is useful to gain feedback on this and other missing elements of the building and the facilities it offers because POE is about seeking the occupants' views of the overall building performance, including its design and operation.

Table 6.1 Topics for feedback and questionnaire

Design and space	Environmental conditions
Desk space	Indoor temperature (winter/summer)
Task chair	Temperature control
Desk screen height	Air movement (winter/summer)
Desk layout, density and configuration	Fresh air ventilation
Personal and team storage	Indoor air quality
Furniture comfort/ergonomics	Odours
Formal meeting rooms	Humidity
Informal meeting areas	Electric lighting
Breakout space	Daylight/natural light
Quiet areas	Glare
Flexibility/adaptability of space	Visual privacy
Signage/wayfinding	Lighting control
Aesthetics/decoration	Acoustic privacy/acoustics
Branding	People noise
Art	Equipment/air-conditioning noise
Plants	External noise

Facilities and support	Technology
Reception	PC/laptop hardware
Conference suite	Mobile devices
Restaurant/café/deli/vending/catering	Software
Tea-points/on-floor kitchen	Network and connectivity
Gym/welfare	Videoconferencing
Cleaning services	Telephony
Security and concierge	Printers and peripherals
Mail/post	Power supplies
Reprographics	Helpdesk and speed of response
Moves/churn	Maintenance and servicing
Helpdesk and speed of response	
Maintenance and servicing	

Base-build	Support of (work) activities
Lifts/elevators – provision, waiting time	Collaboration and teamwork
Stairwells and lobbies	Creativity and innovation
Toilets – provision, quality and gender	Concentration, focus, reading and thinking
Showers and changing rooms	Confidential/private conversation
Landscaping	Communication, knowledge sharing and
External appearance and design	presentations
Accessibility – pedestrian, inclusivity	Contemplation, relaxing and re-energising
Location and transport links	Social interaction
Parking – car, motorbike, cycle	Meetings – informal, formal and one-to-one
Cyclist and runner facilities	Telephone calls and videoconferencing
Local amenities – shops, cafés	PC/laptop processing
	Accommodating/entertaining visitors

(Continued)

Table 6.1 (Continued)

Occupant performance	Organisational factors
Decision making	Management and leadership
Knowledge sharing	Colleagues
Meeting deadlines	Culture
Minimising errors	Career development
Downtime	Reward and motivation
Distraction/concentration	Morale/team spirit
Tiredness/alertness	Autonomy and timekeeping
Work-related stress	Internal communications
Overall performance/productivity	Customer care/satisfaction
Health and wellbeing	**Occupancy patterns**
Stress/burn-out	Time in/out of the office
Fatigue	Time working at home
Symptoms	Time at desk
Absenteeism	Time in meetings
Wellbeing	Time travelling on business
	Remote working policy
Sustainability	**Background variables**
Modes of transport and links	Primary workplace – office/home/co-work
Facilities for bicycle commuters	Desk location – office/open-plan/hot-desk
Journey time	Building location – floor/zone
Proximity to window	Number of people in space/room/office
Ventilation strategy – openable windows	Department/team
Recycling	Job function, role and grade
Renewables and energy management	Tenure – time with the organisation
Biophilia (plants) and biodiversity	Age group, generation and family status
Inclusion and diversity	Gender
	Ethnicity
	Personality profile

Something else to consider is that if a standard questionnaire is used, omitting questions means that it is more difficult to benchmark the questionnaire results against other buildings because each questionnaire may have different missing data. Omitting questions particularly affects "indices" – the overall ratings calculated from the responses to a series of questions, discussed later in the next section.

Questionnaire responses

Response scales

The means by which respondents answer each question is also important. Typically, responses may be as follows.

- **Multiple-choice** – a single response is made from a range of options.
- **Checkbox** – more than one response can be selected from a range of options.
- **Open-ended** – the respondents enter text or data in a comments box.

There are two types of multiple choice scales as follows.

- **Categorical options** – the respondents select one (multiple-choice) or more (checkbox) items from a list of related terms, such as "office", "home" or "elsewhere" in response to the main place or work.
- **Rating scales** – pre-determined numbered choices that represent a judgement of magnitude, value or quantity, for example ranging from "unsatisfactory" (1) through to "satisfactory" (5).

Rating scales are widely used in occupant surveys, and there are various types of scale that can be used, as shown in the table. No specific type of rating scale is better because the scale chosen will depend on the types of questions to be asked and the expected outcomes. In a brief analysis of POE surveys, I found that approximately half used semantic differential scales and the other half used Likert-style labelled scales.

Table 6.2 Common rating scales

Dichotomous scale	Only two choices, such as "agree" or "disagree"
Thurstone scale	Respondents agree or disagree with a series of related statements, a dichotomous precursor to the Likert scale
Labelled rating scale	Multiple-choice answers to a question in the form of a series of labelled/itemised tick boxes, which are usually progressive and symmetrical, and sometimes the points are numbered as well as labelled: Very Dissatisfied Dissatisfied Indifferent Satisfied Very Satisfied O O O O O
Likert scale	A classic labelled rating scale where the respondents rate their level of agreement with a series of statements. Likert scales are often 5-point scales: Strongly Disagree Disagree Neutral Agree Strongly Agree O O O O O
Semantic differential scale	Semantic differential scales offer a range of options with labelling only at either end, where the endpoints are labelled with adjectives of opposite meaning and the options are usually, but not always, numbered: Very Dissatisfied Very Satisfied O O O O O O O 1 2 3 4 5 6 7
Bipolar scale	A scale measuring either positive or negative response to a statement such as "very dissatisfied" to "very satisfied" with a neutral mid-point, rather than a unipolar scale which might be labelled "not at all satisfied" to "completely satisfied"
Rank order scale	The respondents order a range of options according to their preference or evaluation, a ranking number is assigned to each option (usually from a pull-down menu in online surveys)

With increasing use of online surveys, other types of rating scales are used. For example, matrices allow the clustering of related questions and scales. Most online survey platforms, such as SurveyMonkey[1], Qualtrics XM[2] and Google Forms, offer sliding scales, drop-down menus, star ratings, images and icons/emojis (like smiley faces). Icons are more commonly used in surveys aimed at younger people or on short satisfaction surveys, such as those used in cafés and restaurants, or on push-button assessment devices (like those found in toilets or at the exit to airport security). Photographs can also be embedded into questionnaires and on-line surveys to illustrate specific design points or facilities in the building. More visual means of presenting and responding to questions are also available through companies offering bespoke online surveys, such as The Curve AI.

Some researchers, such as Bunn (2020), consider it best practice to use the same type of response scale because "switching between scales in a survey can confuse and alienate respondents", resulting in partially completed surveys. I agree to some extent, but I have found that a range of different rating scales in blocks/pages adds interest and allows similar topics to be asked in different ways, allowing verification. Furthermore, some researchers consider mixed scales more difficult to compare statistically, whereas I have found that they can be compared using the appropriate statistical tests, such as correlations.

Number of points on rating scales

As well as the type of scale, the number of points on the scale is also an important consideration. It will vary with the scope and intent of the survey, but my high-level assessment of POE questionnaires concluded that 5-point and 7-point rating scales appear to be the most popular. However, three-point scales are often used; and nine-point scales are used occasionally. Some researchers prefer an even number of points, with four being the most popular.

There should be sufficient choices to cover a range of responses, but not so many that the distinction between them is blurred. With 5-point scales, the respondent can easily distinguish between the extreme (end) points and the middle point of a scale and can readily judge a further two points midway between the extremes and middle. Labelled/itemised scales tend to use 5-point ratings because it is a challenge to find seven meaningful descriptors of increasing value. For example, a 7-point scale labelled: "very poor", "poor", "*slightly poor*", "okay", "*slightly good*", "good" and "very good" reads less well and is less meaningful than its 5-point equivalent, which would exclude the italicised labels. Seven points are more popular when using semantic differential scales, which assume the respondent will interpret the five unlabelled mid-point scales in a way meaningful to them and that all respondents will interpret the scale in a similar way.

Psychological experiments have shown that people can readily differentiate between seven sounds, colours or odours before making an error, and so seven is regarded as the sensible maximum number of points on a rating scale (Miller, 1956). However, other researchers prefer the simplicity of the three-point scale, which allows the respondent to express a positive, negative or neutral opinion or evaluation. Therefore, as a rule of thumb, limit scales to 5 ± 2 points.

Conventionally, rating scales provide an odd number of alternatives to allow a mid-point "neutral" or "indifferent" option. Rating scales with an even number of options are sometimes referred to as "forced choice" questions because the respondent has to respond positively or negatively and cannot "sit on the fence". Even numbered scales discourage respondents from lazily selecting the mid-point, but sometimes the respondent is genuinely feeling indifferent. Related to this is whether to provide "don't know" or "not applicable" options. On the one hand, such options are an easy opt-out for the respondent, but on the other hand, not allowing them could lead to inaccurate information being provided by the respondent. A "don't know" or "unable to answer" response is not necessarily the same as feeling "indifferent".

Rating scales are often numbered as well as labelled, but not always. When using numbers, some researchers consider a rating of "1" to be better than a "5" – like a podium position with "1" representing first and the winner. However, I prefer to use higher numbers on scales to represent a more positive or more agreeable answer (and it also makes interpreting graphs easier). On bipolar scales with an odd number of response points, some researchers use negative and positive numbers with "0" as the central point. For example, a 5-point scale would range from -2 to +2, which helps to emphasise negative and positive responses.

Indices

An index combines multiple related responses from a set of questions into a single rating or score, for example, by averaging or summing and possibly converting to a percentage of the maximum. Using an index derived from several questions covering the same topic is considered a more reliable measure than using a single question on that topic.

The questions on the original Likert scale, all labelled from "strongly agree" to "strongly disagree", were averaged to produce an agreeable index, what Likert termed a "summative scale". Indices are also commonly used in personality profiling questionnaires, such as the Big Five Personality Inventory (John and Srivastava, 1999), where specific questions are combined to determine the ratings of different personality traits.

Indices are also used in POE surveys to express the overall perception of comfort, satisfaction, productivity or wellbeing. For example, the POE feedback survey I use today was co-created in 1997 (Oseland and Bartlett) with the Office Productivity Network (OPN). We used a common satisfaction rating scale throughout the questionnaire, which means the questions can be combined to form an overall satisfaction index. The index and the rating scales are all converted to percentage satisfaction scores for ease of interpretation. On the other hand, the BUS occupant feedback survey also uses an index of satisfaction, which is derived by combining the series of satisfaction rating scales that are used regularly throughout the survey.

Several established questionnaires have used indices to address productivity. A series of questions in the OPN Survey related to performance were averaged to produce a productivity index. The index was calculated by combining a series of questions asking about the impact of the workplace on specific activities or tasks, including meeting deadlines, creativity and team-working. Other POE professionals, such as ABS Consulting in the 1990s, included questions related to more organisational and cultural issues, such as reward, management and team structure, creating a performance index that they termed the "overall liking score" (OLS).

The development of performance indices is not a simple matter. Although performance indices can illustrate the direction of change in performance, they are not so valid in predicting the actual amount of increase or decrease in productivity. For example, the index depends on the number and type of questions asked, so adding or eliminating certain questions will change the index value. Consequently, it is preferable to use rating scales and indices of self-assessed productivity to estimate the relative change before and after a project.

Some POE questionnaires combine two series of questions to create an index. For example, the OLS asks the respondents to rate the importance of each aspect of the building in addition to how much they like or are satisfied with it. The respondents' rating of importance was then used to weight the rating of satisfaction, and the weighted responses of each aspect of the building can then be added to provide the OLS. In practice, this approach appears to be quite successful because it allows clients to dismiss any factors that cause dissatisfaction but are not strategically important. In contrast, statistical purists would not accept the multiplication of two subjective rating scales.

The more recent Leesman survey produces an overall rating of user experience termed the Leesman Index (Lmi). The Lmi is based on ratings of workplace factors and those considered most important, but the actual calculation of the index is a trade secret.

Some POE methods apply weightings to reflect the importance of the questionnaire items that are combined to make up the index. Basically, the rating of each individual item is multiplied by a preagreed fixed amount, usually a proportion or percentage, so that its contribution to the total score is increased or decreased. The weighted ratings may also be normalised so that the sum of them adds to 100%.

The Net Promoter Score (NPS) is also an index but based on one question only. It uses a scale from 0 to 10 and provides a rating out of 100%. The NPS is used as a measure of customer satisfaction and simply asks customers how likely they would recommend a particular product/service to a friend or colleague. The NPS could be used to as part of a building project evaluation. The NPS is calculated by tallying up the responses and subtracting the percentage of detractors from the percentage of promoters, where promoters are those who responded 9 or 10 on the scale and detractors selected 0 to 6.

Questionnaire design and format

First impressions are important, and a well-formatted questionnaire with well-worded questions is essential for obtaining meaningful responses as well as for improving the response rate.

Well-formatted questionnaire checklist

The following checklist presents some basic principles for the design and format of a questionnaire.

Table 6.3 Checklist for questionnaire formatting

Give the questionnaire a short and meaningful title	☐
Start with a welcome message and introduction, and include instructions on how to complete the questionnaire	☐
Offer the option of anonymity (sometimes it can be useful to have names for follow-up), but reassure people that the data will be presented as averages	☐
Avoid long sentences, long words and acronyms; in other words, keep it short and simple (the KISS principle)	☐
Start the survey with some easy, interesting and non-threatening questions; for example, a few background questions could be asked at the beginning of the questionnaire because they are easy to answer	☐
Place the most important and key questions early in the questionnaire, in case of survey drop-out and push background ones to the end	☐
Keep background questions to a minimum and only ask those which are necessary and help meet the objectives of the POE	☐
Avoid sensitive questions, such as those related to salary and management	☐
Minimise the number of questions asked	☐
Make the questionnaire attractive, easy to read, and spaced out with a consistent layout	☐
Use bold and italics to pick out key words or to mark subtle differences between questions	☐
Organise the transition of questions and group the questions into coherent sections – each question should follow on logically from the previous one rather than as random topics; otherwise, the questionnaire appears disjointed	☐
Where possible, use a similar format for each question, but avoid the risk of respondents falling into a fixed response pattern by using relevant questions only	☐
Allow "don't know" (DK) or "not applicable" (N/A) responses to appropriate questions	☐
Minimise the use of open-ended questions – although they can provide unprompted responses, they can take considerable time to analyse but do provide a space for general comments to be made	☐
In online questionnaires, ensure that the respondent does not have to scroll horizontally	☐
In online surveys, scrolling vertically is more acceptable but minimises the number of questions per page/screen	☐
Similarly, minimise the use of drop-down lists – although they save space, the respondents are less keen on scrolling through possible answers	☐
Test the questionnaire several times before launching it on the web or administering a desk-drop, and ideally, fully pilot the questionnaire	☐

Well-written questions checklist

The wording of the questions is crucial to ensure reliability and validity. The poor translation of questionnaires may also have a detrimental effect on validity and reliability. The following checklist is designed to help identify well-written questions.

Table 6.4 Checklist for writing questions

Ensure clarity – questions must be clear, succinct and unambiguous	☐
Ask for only one piece of information in each question; alternatively, ensure the response options clarify the answer For example, if the response to "do you like the winter and summer indoor temperature?" is "no", then it is not clear whether the respondent does not like just the winter temperature, the summer temperature or neither winter nor summer	☐
Avoid leading questions and do not force or imply a desired answer For example, "do you consider the temperature in your office to be poor?" is a leading question; instead, consider "how do you rate the temperature in your office?"	☐
Similarly, avoid too many questions worded negatively	☐
Do not assume all survey participants are in the same situation, and do not write questions that may exclude some of them For example, "are you satisfied with your laptop?" would exclude those who do not use laptops	☐
Do not assume the survey participants all have the same knowledge For example, "are you content with the amount of spend on maintaining the office temperature?" requires an understanding of the maintenance budget to be answered with any meaning	☐
Do not ask questions with social or judgemental implications because respondents may answer in a way that makes them look or feel better (known as prestige bias or the halo effect)	☐
Write the questions in a non-threatening manner so that respondents are not afraid of the consequences of answering	☐
Avoid asking embarrassing or personal questions and, if required, place any sensitive questions at the end of the questionnaire	☐
Make the questionnaire attractive, easy to read, and spaced out with a consistent layout	☐
Avoid hypothetical questions, as they produce inconsistent information A well-known researcher adage is "hypothetical questions produce hypothetical responses"; for example, "if you were the building manager with an unlimited budget what would you do about the indoor temperature?"	☐
Do not use unfamiliar words, jargon or abbreviations, especially out of context, or be sure that any used are known by all respondents For example, PC may mean "practical completion", "politically correct", "personal computer" or "police constable" depending on the respondent's background	☐
Avoid branching/routing – this is where only certain questions need to be answered depending on the answer to a previous question For example, "if you answered yes to Q1 then answer Q2 and Q3 otherwise go straight to Q4"; however, routing is easier in online surveys or if the questionnaire is administered using a trained interviewer	☐

Question response scale checklist

Like the wording of the actual question, the wording of the response scale needs to be well structured and well formatted.

Table 6.5 Checklist for response scales

Quantify the quantifiers and avoid words such as "very", "usually", "often", "sometimes", "occasionally", "seldom" and "rarely" as they have different meanings for different respondents	☐
Make the response scale exhaustive; that is, the response options or scales should cover all possible answers to a question For example, if the only response options to "where do you usually sit?" are "open-plan desk" and "private office", it would exclude those in a "shared office" or at a "hot-desk" – note that including "other" as a catch-all response ensures all questions are exhaustive but requires more effort to analyse	☐
Provide a full comprehensive range of response options For example, if asking, "how do you rate the temperature in the office?" do not just allow limited responses such as "excellent", "very good" and "good" but also include "unsure", "poor" or "okay"	☐
Ensure the response options are mutually exclusive For example, in response to "where do you usually sit?" do not offer "private office", "open plan" or "on the top floor" as the respondent may be located in a private office on the top floor and therefore not know which option to choose – either two questions are required, or multiple choices allowed	☐
Provide variability or an appropriate range in the response options For example, if the question "what is your opinion of the temperature in the building?" was accompanied by the options "it is the best ever thermal environment I have worked in", "it is between the best and worst ever thermal environment I have worked in" and "it is the worst ever thermal environment I have worked in" then it is likely that most respondents would select the middle option	☐
Do not make the response on the end-points so extreme that respondents rarely use them: this leads to central tendency bias	☐
Avoid asking the respondents to rank-order a series of more than five items	☐
Avoid emotionally loaded adjectives that can have negative connotations, such as "cheap" or "bad"	☐
Minimise the number of questions asked that require the respondent to make calculations: if required, break them down into several questions For example, "what percentage of the time do you spend working in the office, at other corporate locations, at client sites, at home, while travelling and elsewhere?"	☐
Avoid questions where the respondent will need to search for and enter specific data, such as average temperatures or energy consumption	☐
If using emojis, ensure they are relevant and consistent; for example, use: ☹😐☺ to represent dissatisfied, indifferent and satisfied	☐

Piloting the questionnaire

Even if an established questionnaire methodology is adopted, it is likely that some questions will need to be omitted, adapted or added to suit the circumstances. It is, therefore, wise to pilot the questionnaire among a small group to double-check it works well.

For an established questionnaire, the client and project team could complete the survey and make comments, and some online surveys have a review mode. In addition, a new questionnaire should be piloted on a small group of "friendly" occupants from different parts of the organisation. Several iterations may be required before the final wording and layout are approved by the project sponsor.

If a questionnaire is to be translated, in addition to piloting in the new language it should be back-translated to the original language to check for translation errors.

Notes

1 www.surveymonkey.co.uk/
2 www.qualtrics.com/uk/

7 How to conduct a POE – technical performance

Subjective occupant feedback is fundamental to POE, but this chapter expands on the more objective technical methods that mostly provide an independent quantitative measurement of the building's performance. Such methods determine how the building is performing compared with the required conditions and/or those predicted at the design stage.

Space analysis

Space analysis broadly covers the evaluation of how the building space is planned by the designers, along with how the occupants use it. Most POEs, especially those of workplaces, will include elements of space analysis. Space planning efficiency appears to be more popular in post-project evaluations, whereas space utilisation is more commonly measured pre-project to inform the space requirements of the new space.

Space planning metrics

Measurement and analysis of the building plans can be used to generate space metrics that express how efficiently the space in the building will be used. A typical space analysis will require the following measurements to be collected to generate popular reported space metrics (see Figure 7.1). Professional bodies around the world, such as the RICS in the UK, Building Owners and Managers Association (BOMA) in the US and the Nederlandse Norm (NEN), provide guidance on standard ways of measuring space in buildings.

- **Gross internal area (GIA)** – The area of a building measured to the internal face of the perimeter walls at each floor level, including the core areas.
- **Net internal area (NIA)** – The area within a building measured to the internal face of the perimeter walls at each floor level, excluding core areas (see below).
- **Core areas** – These are the common or shared areas outside of the main NIA. Toilets, stairwells, lobbies, elevators (lifts), plant rooms, columns, vertical ducts and corridors linking leased spaces are all part of the core. What is included as part of the core varies across countries.

DOI: 10.1201/9781003350798-7

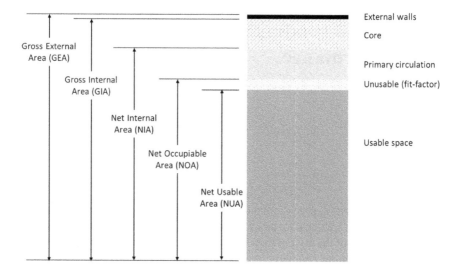

Figure 7.1 Building space metrics

- **Net lettable area (NLA)** – The NLA (sometimes known as the net rentable area; NRA) refers to the space leased/rented. It is often similar to NIA, but because what is included as part of the core varies from country to country, the NIA, NLA and NRA may also differ.
- **Primary circulation** – The legally required main circulation routes connecting the building cores, particularly the elevators, stairwells and toilets. Local planning legislation dictates the distance and width of the primary circulation routes.
- **Secondary circulation** – The aisles between individual desks and other spaces.
- **Net occupiable area (NOA)** – According to BCO, the NOA is the NIA minus the primary circulation routes.
- **Unusable space** – Any areas within the NIA that cannot be easily occupied by a person (or more specifically in the workplace by a desk), including space under stairwells or in tight angular corners. The unusable space is used to calculate the fit-factor and may actually be referred to as the fit-factor.
- **Net usable area (NUA)** – For most workplace consultants, the NUA is the NIA minus the primary circulation routes and unusable space. However, others consider the NUA to be the same as the NIA or NLA.
- **Desk count** – Traditionally, this would have included the number of open-plan desks and private offices. In agile and activity-based working environments, it may include alternative spaces that can be used as a workstation, such as touch-down tables and hot-desks, quiet pods, focus booths and breakout spaces. The term work-point is sometimes used to capture both the more traditional desks and alternatives.
- **Headcount** – The number of occupants now tends to be different to the number of desks, so the headcount will need to be obtained separately. Bear in mind that in workplaces, the accommodated (or assigned) headcount provided may

or may not include contractors and temporary staff, and it may refer to those employees who only occasionally use the building. For other building types, the headcount may include more transient building users, such as students in education settings, patients in healthcare and delegates in conference centres. Sometimes the number of people is reported as the full-time equivalent (FTE) rather than the actual number of people in order to account for those who are not working full-time and so only use the building at certain times during the week or year.

- **Support space** – The area of support space on a typical floor or across the whole building includes meeting rooms and informal meeting areas, plus staff amenities such as tea points, breakout areas, a restaurant, a gym or a prayer room.
- **Meeting rooms** – The number of meeting rooms, and the associated number of seats, or the amount of space taken up by meeting rooms, or meeting areas, may be measured.
- **Enclosed office space** – The area occupied by enclosed private and shared spaces rather than the open-plan areas.

The above measurements are used for calculating the key space planning metrics. The following metrics are usually reported in a POE.

- **Spatial density** – This is the most common space metric reported in POEs, sometimes referred to as the workplace density or occupancy ratio, cited as the m² per person. The person/people denominator in the ratio may refer to the regular building occupants or possibly to the total headcount, including occasional users of the building. The density may also be expressed as the m² per desk: this is sometimes referred to as the static density because the number of desks is fixed. In contrast, in an agile working environment the m² per headcount is sometimes called the dynamic density because the actual occupancy changes over time. To confuse matters further, the number of people may be converted to FTE to capture part-time employees. Furthermore, the density may refer to the whole building or just part of it, such as the main office floors. The key message here is to report density but, as it is not as simple as it appears, first agree on how to collate the data and then ensure that when comparing with other buildings, or benchmarking, the same density metrics are adopted. For example, the BCO (Oseland et al., 2022) recommends a special density of 10–12 m² per person for a typical office floor, where the "per person" refers to the planned number of occupants. Less used in POE but more commonly quoted in research is social density, the number of people per given area of space (or m²).
- **Plan efficiency** – Space planning metrics succinctly highlight the space efficiency of the building. For example, plan (or landlord) efficiency is the ratio of NIA to GIA expressed as a percentage. The BCO (2014) recommends a plan efficiency of 80–85% to ensure there are adequate but not superfluous core areas.
- **Tenant efficiency** – According to the BCO (2014), primary circulation covering around 15–22% of the NIA in offices is considered efficient. However, the ratio varies according to the building function and numbers of people moving around it. The converse ratio of NOA to NIA, sometimes referred to as the tenant (user

or occupier) efficiency ratio, should therefore be 78–85%. The ratio of NUA to NIA may also be used.

- **Fit-factor** – This is the ratio of unusable space to the NIA, expressed as a percentage. It provides an indication of any space inefficiencies due to the shape and structure of the building. Ideally, the fit-factor should be 5% or lower but it could be higher in legacy buildings compared with modern ones that have more open contiguous space.

In addition to the above, I include the following space metrics in a POE of offices.

- **Support space** – The spatial density is a good indicator of space efficiency. However, while a high density with less space per person may be considered "efficient", it could simply indicate a lack of facilities. For example, in an office, more desks could be packed into the building in place of support spaces, such as meeting rooms or employee amenities (like kitchen areas and breakout spaces). Any report of spatial density should therefore be accompanied by a measure of the level of support space (see Figure 7.2).
- **Meeting room ratios** – Offices are increasingly being used for collaboration. The number of meeting rooms and the capacity of those rooms needs to be matched to the occupants' requirements. The ratio of meeting rooms and meeting room seats to the number of occupants (desks) is a useful metric. The BCO (2011) provided some guidance on the ratio of different sizes of meeting rooms required for different market sectors, which could act as a benchmark. For example, for the public sector, the BCO proposed one open-plan meeting space per 20 staff, one 1–2 person quiet room per 40 staff, one 3–4 person meeting room per 60 staff, one 6–8 person meeting room per 100 staff, one 10–12 person meeting room per 150 staff, one 12+ person meeting room per 200 staff and one 20–25 person meeting room per 400 staff.
- **Enclosed office space** – This metric is more useful in traditional office layouts than in modern open-plan spaces.

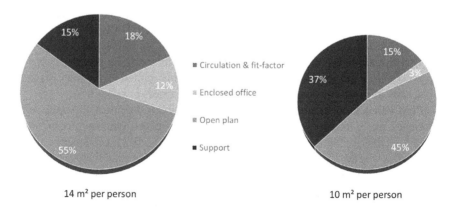

Figure 7.2 Example space analysis of a traditional (left) versus a modern (right) office floor

- **Sharing ratios** – In more agile and activity-based working environments, it is useful to understand the ratio of desks to the assigned headcount. The ratio may be expressed as either the headcount per desk or desks per headcount, and the latter may be converted to a percentage. For example, 10 employees per 8 desks would be 10:8 or 1.25:1 staff per desk, or 8:10 desks per staff or 80%.

The space analysis may also involve a more qualitative evaluation to capture quality and functionality. For example, the building location, size of contiguous space, local amenities, sub-divisibility, depth of building (and daylight ingress), accessibility (including a dedicated or shared entrance), services and infrastructure are all key aspects of a building that affect how it supports the occupants. These variables can also be assessed using the building plans as part of a building appraisal prior to occupancy.

Utilisation studies

Space analysis determines how efficiently the space is planned, but it does not reveal how efficiently the space is used over time. For example, a high-density space that is only occupied for short periods of time, such as a staff restaurant, may not be considered space efficient.

A space utilisation survey, also known as a time utilisation study or occupancy analysis, involves monitoring how different spaces within the building are occupied across the day and week. Depending on the method deployed, the monitoring may be hourly or in real-time, and it may be conducted over one or two weeks, over several months or indefinitely. Such studies are more prevalent in offices and higher education institutions that audit a range of spaces, including open-plan desks, private offices, meeting rooms, informal meeting areas, breakout spaces and staff restaurants.

At the most basic level, it is recorded whether a space is simply occupied or unoccupied (vacant/empty). Depending on the system used to collate the data, the space may also be denoted as temporarily unoccupied (temp. unocc.), sometimes called "signs of life", when the space looks like it may be in use because there are personal belongings present (such as a jacket, bag or laptop) although there is no one physically present. In desk-sharing (hot-desking) environments, some occupants may set up their belongings on a desk at the start of the day but then spend much of their time in other spaces, such as meeting rooms – a method of reserving a desk that is colloquially termed "beach-towelling".

Utilisation usually relates to the frequency of use; that is the time that the space is occupied as a percentage of the time the space is available. The time available is usually considered to be 40 hours per week (from 9 am until 5 pm, Monday to Friday) but can vary depending on the type of building and space. For example, Figure 7.3 shows the hourly utilisation of all desks across the working week. Utilisation is usually considered to be the percentage of occupied plus temporarily unoccupied desks.

If a group space such as a meeting room is monitored, then the number of people present may also be counted and compared with the capacity of the space. Sometimes referred to as "occupancy", it is reported by calculating the number of people present as a percentage of the room capacity. Confusingly, in studies of rooms in

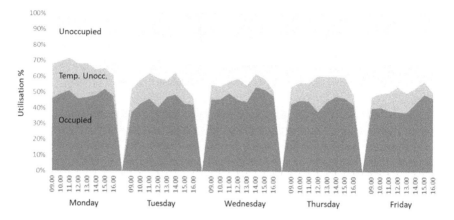

Figure 7.3 Office desk utilisation across the working week (pre-pandemic)

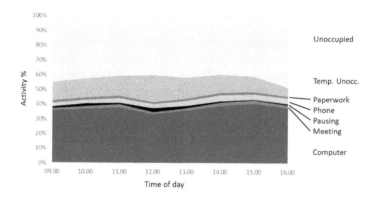

Figure 7.4 Office activity averaged across the day

higher education institutions, the percentage utilisation is calculated by multiply-ing the percentage frequency by the percentage occupancy.

A variation of the utilisation study is a task observation study, where data is also collected on the activity the occupant is performing while in the space, such as computing, reading/writing/paperwork, telephone usage, formal meeting, infor-mal meeting/chatting and pausing/thinking/break. Task observation studies may be used to determine the spaces required to support occupants' work activities, for example see Figure 7.4.

Measuring the utilisation of, say, a group of desks may appear to be a straightfor-ward task, but it is essential to understand how the utilisation figure is truly calculated.

The utilisation figure may be a simple average of the utilisation of each space. However, in most cases the overall, or average, utilisation figure usually repre-sents the percentage of desk-hours occupied, that is the number of desks occupied at each hour of all the hours available. For example, if there are 100 desks and a

Desk	Time of day							
	9.00	10.00	11.00	12.00	13.00	14.00	15.00	16.00
1	O	O	O	O	O	O	O	O
2	O	O	O	O	O	O	O	O
3	O	O	O	O	O	O	O	O
4	O	O	O	O	O	O	O	O
5	O	O	O	O	O	O	O	O
6	U	U	U	U	U	U	U	U
7	U	U	U	U	U	U	U	U
8	U	U	U	U	U	U	U	U
9	U	U	U	U	U	U	U	U
10	U	U	U	U	U	U	U	U

Scenario A

Desk	Time of day							
	9.00	10.00	11.00	12.00	13.00	14.00	15.00	16.00
1	O	O	O	O	U	U	U	U
2	O	O	O	O	U	U	U	U
3	O	O	O	O	U	U	U	U
4	O	O	O	O	U	U	U	U
5	O	O	O	O	U	U	U	U
6	U	U	U	U	O	O	O	O
7	U	U	U	U	O	O	O	O
8	U	U	U	U	O	O	O	O
9	U	U	U	U	O	O	O	O
10	U	U	U	U	O	O	O	O

Scenario B

Desk	Time of day							
	9.00	10.00	11.00	12.00	13.00	14.00	15.00	16.00
1	U	U	O	O	O	O	U	U
2	U	U	O	O	O	O	U	U
3	U	U	O	O	O	O	U	U
4	U	U	O	O	O	O	U	U
5	U	U	O	O	O	O	U	U
6	U	U	O	O	O	O	U	U
7	U	U	O	O	O	O	U	U
8	U	U	O	O	O	O	U	U
9	U	U	O	O	O	O	U	U
10	U	U	O	O	O	O	U	U

Scenario C

Desk	Time of day							
	9.00	10.00	11.00	12.00	13.00	14.00	15.00	16.00
1	U	U	U	U	O	O	O	O
2	O	O	O	O	U	U	U	U
3	O	U	O	U	U	O	O	O
4	U	U	U	O	O	O	O	U
5	O	O	O	O	O	O	O	O
6	U	U	U	U	U	U	U	U
7	O	O	U	U	U	U	O	O
8	U	U	U	U	U	O	O	O
9	O	O	O	U	U	U	U	O
10	U	U	U	O	O	O	O	U

Scenario D

Figure 7.5 Four scenarios illustrating 50% desk utilisation over one day (O = occupied; U = unoccupied)

40-hour week then there are 4,000 desk-hours available. A utilisation of 50% could therefore be due to the same 50 desks being in use full-time, and 50 unoccupied all the time, or it could represent 50 different desks being used for half the time, or a mix of different desks occupied for different lengths of time. The simplified diagrams in Figure 7.5 show four different scenarios of 50% utilisation for 10 desks occupied and unoccupied over the course of one day. If the objective is to maximise the utilisation with the least number of desks, as in an agile/hot-desking environment, then scenarios A and B will be easier to manage.

Therefore, rather than just the overall or average utilisation, it is worthwhile calculating the peak utilisation, such as the maximum proportion of spaces in use at any one time or the spaces occupied for a whole day versus those only occupied for half a day or fully unoccupied for a day, and so on. Different utilisation studies will use different calculations, so it is essential to agree on the metrics in advance, especially for comparing new results with previous surveys or benchmarking against others.

The information provided by utilisation studies can be collated using different methods, each with its own advantages and disadvantages, as follows.

- **Observers** – The longest-standing method for collating utilisation data is to use trained observers. Originally the observers used printed space plans to record their observations, but today mobile devices are used. This means that the data does not need to be transposed and has less chance of being lost because it is captured in real-time or, at minimum saved at the end of each day. The observers will visit the spaces hourly, typically between 9 am and 5 pm, and record the occupancy status. They can also record the number of people present in a space and the activities taking place. The number of observers required depends on the number of spaces and their accessibility. Depending on the observers' ability, and the complexity of the space, they can each record from 100 to 200+ spaces per hour.

- **Building sensors** – There are several types of occupancy sensors on the market for capturing utilisation data, but they are broadly categorised as under-desk or overhead. Desk sensors are placed on the underside of desks or meeting tables to capture whether the seat is occupied or unoccupied. Most use passive infrared technology to detect a slight difference in heat (due to a "hot body"), then send an instant update providing real-time data recording. Ceiling- or wall-mounted sensors may also use passive infrared sensors to detect motion, which is particularly useful to determine people entering a room, including those who, say, just take a quick call in a meeting room but do not sit down. More sophisticated overhead sensors are the people counters, using a mix of infra-red light beams, stereo vision cameras and "time of flight" 3D cameras combined with AI to detect the number of people in a space. Sensors may be linked to the existing building systems, such as the BMS or meeting room and desk booking platforms.
- **Booking systems** – A room or desk booking system will provide easily accessible high-level information on the anticipated occupancy levels. Unfortunately, there is often a discrepancy between the spaces booked and how they are actually used. Booking systems that do not have a check-in to confirm the space is in use may provide misleading information. Furthermore, they can make spaces appear busier than they really are, discouraging occupancy.
- **Card access control** – If the building is fitted with an access control system (like swipe or proximity card access), it is possible to use such data to determine the number of people in the building over time. This method provides a high-level indicator over long periods of time, and the data is usually collated without the need to commission an additional survey. The downside is that the data is often difficult to unravel, it is subject to error (for example, because of tailgating), and it does not provide information on which specific spaces (like desks) are used. However, it is worthwhile comparing such data with observation studies to calculate the adjustment required to provide more accurate figures.
- **Computer activity** – Computer (sometimes termed "employee") monitoring or tracking software can be used to capture when a computer processor is active. In the past, such software was used by some organisations as an indicator of desk occupancy. However, the process relied on marrying up the unique reference (its IP address) of a single computer to a specific desk, which worked for fixed desktop computers but not laptops used in different spaces throughout the building and elsewhere. Also, of course, employees do not spend all of their work time operating their computers.

- **People sensors** – In the past, some researchers have attached wearable sensors to people to track their movement within a building and capture interactions with colleagues. Traditionally, the sensors were quite cumbersome and visible, so they were not useful for discrete studies. However, improvements in wearable technology may lead to more convenient people trackers.
- **Mobile phone tracking** – The global positioning system (GPS) capability of mobile phones can be used to track movement and position. Historically, GPS was used in social research at the city level rather than inside buildings. This is partly due to the accuracy of the GPS, which indoors may be around 10 m, both horizontally and vertically, so it was particularly difficult to use for tracking occupancy in multistorey buildings. However, the technology and its accuracy are continually improving, and indoor positioning systems are now available to track a person's location in a building using their mobile phone.

The advantage of using observers is that they can record additional information about occupancy, such as the activity and equipment used. Observation studies tend to last one to two weeks, and for these shorter surveys, they are less expensive than installing sensors. However, if longer periods of time are required, then observation studies become less practical and more expensive. Sensors become more cost-effective for surveys longer than three to six months, particularly for ongoing monitoring and space management.

The main downside to observation studies is finding sufficient observers for the survey period, then training and motivating them. Observer reliability also varies, with data potentially being lost due to illness or travel issues, although this can be managed by providing reserve observers. In addition, regular spot checks carried out by supervisors will be required to ensure validity and resolve any issues arising. Unlike a sensor-based approach, when using observers, each space tends to be recorded on an hourly cycle rather than in real-time. Amazingly in one of the first utilisation studies I managed, when the occupants noticed the observes nearby, they moved on to another desk so that the space was recorded as more utilised.

Observers peering at desks and into rooms may be considered more invasive than using hidden sensors. However, building occupants often seem more suspicious of sensors, sometimes mistakenly assuming they are cameras and a privacy infringement. Likewise, many occupants are hesitant to give permission for their phones to be used to track them, even though many phone apps access the GPS for location tracking. Occupancy sensors and observers do not identify the specific people present, but phones and access control cards could be more readily linked to individuals. Before choosing the method for measuring utilisation, consider the cost, disruption, access to data, format of the data and privacy concerns. Regardless of the methodology adopted, the purpose of the study, process and data-handling needs to be clearly communicated with the building occupants.

Observation of movement and interaction

While not a common POE method, trained specialists can observe in detail the way that the occupants use the building. Ethnography is an established methodology used in anthropology and involves the observer embedding themselves in a social situation and then studying the behaviour of those around them. Whitemyer (2006) noted that "a POE questionnaire might ask an office worker how many times a day he [*sic*] gets up and walks to the photocopier, informing a designer on carpet durability, desk placement or proper lighting. But an ethnographer will watch people walk to the photocopier, consider how they navigate through the office and observe how they interact with their co-workers".

The approach is mostly qualitative, but specific behaviours such as group interactions can be tallied. Ethnography enhances more traditional POE methods and provides valuable insight into how spaces are used, but unfortunately it can be a costly process.

Sailer et al. (2010) proposed that flow of movement, social network analysis and space syntax may be useful in POEs. For example, observers strategically placed around a building can record the flow of movement through it, tracing the movement onto the space plans (see Figure 7.6). The analysis can be used to highlight primary circulation routes, accessibility of spaces and under-utilised space, and it also may indicate the key (popular) personnel within an organisation. Like most observation studies, the method can be considered intrusive.

Social network analysis is a technique used in sociology to identify how different people (nodes) are linked through social connections such as friendship, family or financial ties. In the workplace, it can be used to gauge the level of interaction between colleagues, which is key to knowledge sharing. The data may be collated by asking about key interactions or analysing information such as telephone and email conversations.

Space syntax can be used to quantify how navigable a space is – particularly useful for buildings where wayfinding is important. It is a process for analysing spatial configurations to demonstrate the relative connectivity and integration of spaces. Space syntax was developed in the 1980s by Hillier and Hanson (1984)

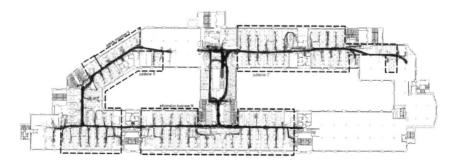

Figure 7.6 Flow of movement analysis (Sailer, 2010)

and has since evolved to help predict the relationship between spatial layouts and behaviour, such as connection and interaction. It is usually used as a predictive tool, but it could also be used to illustrate changes and improvements to a space.

Storage audits

Prior to the advent of electronic filing and storage in offices, a relatively high proportion of the floorplate was taken up with paper files, stored in cabinets, tambour units and high-density roller racking and so on. A measure of the paper filing by the survey team, usually made in linear metres (lm), was common practice and a POE might report a reduction (increased efficiency) in filing. Some POE questionnaires also asked the respondents to estimate or measure their filing.

Indoor environmental quality assessment

More detailed and forensic POEs are likely to include measurements of the indoor environment, gathering data on physical conditions such as temperature, ventilation rates, air quality, acoustics and light. Because of the time, expertise and expense involved, such investigations tend to be conducted if a specific issue with the environmental conditions is revealed during an initial evaluation. The measurements may be used to corroborate feedback, from the occupants or project team, or to compare the modelled and predicted environmental conditions with those achieved.

In my previous book, *Beyond the Workplace Zoo*, I explain how the perception of most environmental conditions only correlates weakly with the corresponding physical measurements. This is due to psychophysics – a person's response to the environment is affected by psychological and physiological factors, including perception, cognition, attitude, personality and age, and situational factors such as activity. Consequently, people sense noise rather than sound level and thermal comfort rather than temperature and so on. Furthermore, there are different measurements of the same environmental parameters, for example air, radiant and operative temperature. Indeed, what is considered the most appropriate physical measurement is regularly debated among engineers, acousticians and physicists.

Physical conditions usually vary throughout a building and across the day. Therefore, another issue is where and when to make the measurements, especially with limited resources (equipment and cost). Different equipment may be used for spot measurements (Figure 7.7) or continuous monitoring: more advanced instrumentation tends to be moved around the building whereas simple data loggers and sensors are usually left in situ over time. A few researchers have developed environmental monitoring instrument carts/trolleys that monitor a raft of environmental conditions while located adjacent to an occupant. Examples include the UC Berkeley team (Schiller et al., 1988) and The Center for Building Performance and Diagnostics at Carnegie Mellon (Loftness et al., 2009). I recall once designing and constructing small cases containing all the sensors for measuring thermal comfort, and asking my research participants to carry them (Oseland, 1995).

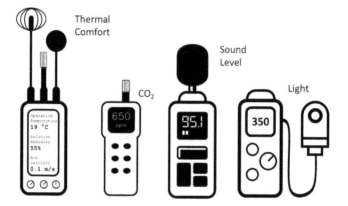

Figure 7.7 Handheld environmental monitors

The indoor environmental monitoring process will need to be discussed in advance, with the indoor factors and metrics plus the type of equipment/sensor, measurement points and timescale all agreed upon in advance. International standards provide best practice guidance on how to make these technical measurements in a consistent way. The guidance is often detailed and complex but should be adhered to, especially for comparing buildings, using benchmarks or being quoted in academic research.

Thermal comfort

International and national standards and guidance on thermal comfort, such as *ISO 7730*, *ASHRAE 55* and *CIBSE Guide A*, refer to operative temperature, air temperature, mean radiant temperature and effective temperature. Operative temperature, also termed resultant or globe temperature, is a combination of air temperature and mean radiant temperature (heat radiated from surfaces) and is considered the one most aligned to the human sensation of temperature. In *Beyond the Workplace Zoo*, I recall that "I was once called in to investigate why staff were feeling cold, and wearing scarves and jackets, in their office when the air temperature was measured at 20°C. The office was located underground in the middle of a large archive and I measured the radiant temperature to be 14°C so the staff were experiencing 17°C operative temperature, which is too low for sedentary activity".

In addition to temperature, *ISO 7730* incorporates relative humidity, air velocity, clothing (clo) and activity (met). These combined parameters are used to predict the mean vote of a group of building occupants on a thermal sensation scale. So, if looking for a relationship between temperature and occupant comfort in a POE, I recommend using operative temperature or predicted mean vote (PMV) rather than air temperature. Nevertheless, although reasonable correlations can be found between perceived and measured temperature, the relationship is unlikely to be as strong as expected due to psychophysics.

Thermal modelling can be used to create a 3D computer simulation of a building to predict how the temperature will vary throughout the space over time. The model takes account of the orientation of the building, local outdoor conditions, building services and internal gains, along with the positions, materials and thermal properties (U-values) of building elements such as windows. The model can predict heating and cooling loads and the associated energy use, usually illustrated as a heat map, and the predicted indoor temperature can be compared with measured values.

Temperature meters tend to use thermocouples for quantifying the ambient temperature, but thermal imaging using infrared cameras might also be used to illustrate the temperature distribution throughout a building. Researchers, including Stevenson (2019), recommend thermography is used when conducting POEs of houses.

Indoor air quality

IAQ refers to the level of pollutants in the air, including volatile organic compounds (VOCs) and (CO_2). Maximum indoor CO_2 levels (measured in parts per million; ppm) and/or the ventilation rates (measured in l/s/person) required to remove pollutants are recommended in standards and guidance like *ASHRAE 62, BS 13779* and *CIBSE Guide A*.

CO_2 is regularly used as a proxy measure of air quality because it is a common pollutant. Humans exhale CO_2 and are the biggest producer of it in offices; thus, by maintaining low CO_2 levels other pollutants are likely to be reduced. There are many basic CO_2 sensors and monitors on the market and a variety of types, including non-dispersive infrared, photoacoustic, electrochemical, semiconductor and catalytic combustion CO_2 sensors. CO_2 monitors are often placed in classrooms, and CO_2 sensors may be linked to the BMS to control ventilation rates. High CO_2 levels may be perceived as stuffiness, warmer temperatures or causing drowsiness.

VOCs are chemicals used and produced in the manufacture of paints, solvents, pharmaceuticals and refrigerants. VOCs are released (off-gassed) by some furniture and building materials, and some are linked with poor health. Many countries, therefore, have regulations to limit VOC emissions from commercial products. Very high VOC levels may result in eye and nose irritation, headaches, dizziness and fatigue. There are thousands of VOCs, so the total volatile organic compound (TVOC) is usually measured as it is easier and less expensive than measuring individual VOCs. There are different means of measuring TVOC with varying degrees of accuracy, complexity and expense. Some handheld air quality monitoring equipment uses photoionisation detection to provide a quick reading.

Like other environmental measurements, investigating the impact of IAQ on occupant feedback, wellbeing and performance requires training and expertise plus the appropriate equipment. Nonetheless, basic CO_2 sensors and data loggers could initially be placed around the building and used as a high-level indicator of IAQ if it is flagged as a major concern of the occupants.

Acoustics

Noise is a major concern of office workers, especially those in high-density open-plan working environments with little screening or separation. *ISO 22955, BS 8233:2014, CIBSE Guide A*, WELL and the ASHRAE *Handbook for Workplace Acoustic Quality Assessment*, to name just a few, all provide guidance on acoustics in a range of buildings. These guides refer to many acoustic metrics, including sound pressure level (SPL), noise rating (NR), reverberation time (RT), speech transmission index (STI), speech intelligibility index (SII), speech clarity (C50), acoustic attenuation of speech (DA,S) and spatial decay rate of speech (D2S). Acousticians appear to be in continuous debate over the most relevant acoustic metrics.

Even the most basic metric of SPL may be measured differently because (1) several weightings may be applied to reflect the sensitivity of the human ear; (2) the sound measurement may be averaged over time, termed the equivalent continuous sound pressure level, or the maximum or peaks reported; and (3) the sound may be measured at different distances from the source. This all adds to the confusion of measuring and reporting acoustics in a POE, especially if making comparisons or benchmarking.

My extensive literature review of noise (Oseland and Hodsman, 2017) revealed that typically just 25% of perceived noise annoyance is attributed to sound level. This may partly be due to the type of sound level recorded but is more likely related to psychoacoustics. Noise is defined as unwanted noise, where the individual decides whether the sound is considered to be noise. The same sound source, and corresponding sound level, may or may not be considered noise depending on factors such as context (time of day, source and type of sound etc.), activity, perceived control and personal factors (such as personality and age).

Monitoring and interpreting acoustics is becoming increasingly complicated. However, there is a range of instruments available, ranging from simple compact sensors, that can be located throughout a building, to high-quality sound level meters. As with all environmental measurements, the methodology, metrics, accuracy and cost need to be agreed in advance.

Light

Both electric lighting and daylight ingress will affect the comfort, wellbeing and performance of a building's occupants. Local legislation usually dictates the electric lighting requirements from a health and safety perspective and guides such as *European Standard EN12464, CIBSE Guide A*, the BCO's *Guide to Specification* and WELL provide best practice advice.

Surface illuminance – either desk, wall or ceiling – is the main metric reported in lighting guidance. Illuminance can be measured using readily available light or lux meters. The sensor is laid flat on the surface to be measured, but it may be necessary to subtract the ambient lighting (measured with the lights off) if calculating the illuminance provided by the electric lighting. Several spot measurements will be required to build up a picture of the lighting across a room or a single representative measurement point should be agreed upon.

WELL, BREEAM and LEED also encompass requirements for daylight in offices, but they refer to the distance of a desk from a window. *CIBSE Guide A* refers to the daylight factor, but predicting the daylight factor in an office space is complex and is usually determined through computer modelling and simulation. Electric lighting that more closely mimics natural light many be considered better. Detailed studies may therefore investigate the colour spectrum range and colour temperature of the electric lighting.

Sustainability audit

With growing concerns over carbon emissions from buildings and the introduction of national (or corporate) net zero carbon targets, there is growing interest in how, from a sustainability perspective, buildings perform compared with how they were predicted to perform. A POE may therefore include measurement and benchmarking of the energy consumed, water used, and waste produced, for example as shown in Figure 7.8.

To assess sustainability more broadly, environmental assessment methods such as BREEAM or LEED might be used to calculate their corresponding environmental rating. This would allow a building to be evaluated pre- and post-project or compared with other rated buildings. However, such assessments are usually carried out by trained and accredited assessors.

Energy, water and waste may be measured differently, so some professional bodies provide guidance on how to measure them to allow for accurate comparison. For example, the Investment Property Databank (IPD) offered guidance on measurement in its now-dated *IPD Environment Code* (2008) and *Global Estate Measurement Code for Occupiers* (2013), and also provided benchmarks for UK offices. In Australia, NABERS provides benchmarks for energy, water and waste. However, as best practice benchmarks change over time, and because they are dependent on some assumed weather conditions and so on, it is better to compare the energy and water consumption and waste production pre- and post-project or across an organisation's portfolio for the same time of year.

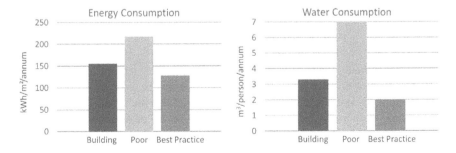

Figure 7.8 Example of energy and water consumption benchmarking

Note that environmental metrics may be reported as consumption per m^2 of floor space or per person/employee/FTE and usually per annum. The metrics used, measurement time period and reporting will need to be agreed upon in advance of conducting the POE.

Energy consumption

In the UK, CIBSE produced guidance on how to measure energy consumption in its technical manual *TM22: Energy Assessment and Reporting Methodology,* and this is recognised as the preferred methodology by professional engineers. Typically, the building's energy use is captured through metering and reported as kilowatt-hours (kWh) consumed, usually the kWh per m^2 or sometimes the kWh per employee in offices. If possible, energy consumption may be broken down by the source (electricity, gas, oil and renewables). The energy consumption may also be converted to CO_2 emissions or energy costs.

Energy data is usually compiled monthly, but comparisons need to be made for similar periods of time to account for seasons and occupancy. Ideally, the data would be adjusted for outdoor temperatures and any other weather anomalies using a system of "degree days". Likewise, energy models draw on databases of the typical weather conditions for the location of the buildings being modelled, so adjustments for any atypical conditions would need to be made. With energy data, any unusual occupancy patterns, such as absenteeism due to viruses, should also be accounted for.

Energy data can be compared to published benchmarks such as the UK government's now-defunct ECON 19 (DETR, 2003) and more recent energy performance certificates[1], and CIBSE's *TM46 Energy Benchmarks* (2008) and its online *Energy Benchmarking Tool.* Other countries also provide energy benchmarks; for example, the ENERGY STAR *Portfolio Manager* can be used to benchmark commercial buildings in the US, and in Australia, the Australian Energy Regulator publishes benchmarks for residential buildings allowing householders to compare their usage against other homes in their postcode.

A building's energy performance is also dependent upon its air tightness, included in some POEs. CIBSE's *TM23 Testing Buildings for Air Leakage* (2022) describes the fan pressurisation and low-pressure pulse methods for measuring air tightness.

Water usage

A building's water usage can also be collated through metering and then compared. The *IPD Environment Code* categorises water consumption as the mains water supplied, water extracted on-site, harvested rainwater and recycled water (greywater). Water is usually reported as the m^3 per m^2 of occupied space, or per employee per year in commercial buildings.

Though less common than energy consumption, water usage benchmarks are available worldwide. For example, the UK's Environment Agency, Singapore's

National Water Agency and ENERGY STAR's *Portfolio Manager* in the US all list benchmarks for a whole range of building uses, including offices, schools, hotels, prisons and libraries.

Waste production

The annual mass of waste arising from a building and sent to landfill and incineration may also be monitored on-site. Most organisations have a record of their waste production because they pay by the tonne for it to be removed. The *IPD Environment Code* recommends identifying the total non-recycled waste and the general waste sent to landfill. In offices, waste may be reported as either the kg or tonne per m² or per employee. As well as the ongoing operational waste, any waste produced during building construction may also be monitored.

Benchmarking waste production is a less common practice than for energy and water consumption, but some specialist consultancies and waste management companies do offer a commercial waste benchmarking service. Nevertheless, it is better to monitor and compare waste pre- and post-project or with other buildings within the organisation's portfolio.

Project management metrics

Project managers will undoubtedly use cost, and completing the project within budget, as part of their evaluation of a project's success. However, they also refer to delivering the project on time and to a high quality. Good project management involves managing and balancing cost, time and quality (Figure 7.9). Time and cost are relatively easy to quantify, whereas quality will include an assessment by the project team and/or building users. As part of a POE, project managers may also report on other factors affecting the project delivery, such as the project team briefing process, procurement process, sharing of information, decision making and resolution of issues. Views on the project delivery are usually collated through a facilitated focus group with the project team.

Figure 7.9 The project management triangle

Cost analysis

The project manager's, and often the occupier's, most fundamental metric for assessing the performance of a building project is cost. The final cost may be compared with that budgeted and/or assessed against regularly published industry benchmarks, such as Statista.[2]

The most common cost metric is the capital expenditure (CapEx) required to complete the project. Professional bodies such as RICS provide guidance on how to capture and benchmark project costs: *Cost Analysis and Benchmarking* (RICS, 2013). For a major building project, the costs may be broken down as shown below.

- **Facilitating works** – including demolition and groundworks.
- **Substructure** – such as excavation.
- **Internal finishes** – to walls, partitions, floors and ceilings.
- **Fittings, furnishings and equipment** – including door-fittings, kitchens, shelving and planting.
- **Building services** – including mechanical, electrical, sanitary, water, air conditioning, lighting and control systems.
- **Work on existing buildings** – such as repairs and renovation.
- **External works** – landscaping, site clearance, paths and roads.
- **Fees** – main contractor preliminaries, overhead and profit, project and design team fees.
- **Additional costs** – other project costs and budget risks (contingency).

In addition, all the furniture and technology costs, which are usually budgeted separately, will need to be added. For an office, the furniture includes all the desks and task chairs plus the meeting room furniture and "loose furniture", such as that used in informal meeting and breakout areas. Technology will include not only personal computer equipment and telephony but also the IT infrastructure.

Capturing all project-related costs is not as simple as it may first appear, but fortunately most building projects will include a cost consultant or quantity surveyor on the project team. With the client's permission, these professionals will be able to hand over an end-of-project account showing all the related costs.

In addition to the project construction costs, a longer-term view of the building performance may include operational expenditure including rent and rates, services and facilities, utilities and property maintenance. For example, Lambert, Smith and Hampton provide a cost benchmarking service and in its Total Office Cost Survey[3] the annual building costs are categorised as below.

- **Property costs** – such as rent and rates.
- **Fitting-out and furniture** – project costs amortised over five or ten years.
- **Hard facilities management** – including insurance security, cleaning, waste, energy and water, plus associated mechanical and electrical repairs and maintenance.

- **Soft facilities management** – telephony, catering, reception, mail and reprographics.
- **Building management** – associated fees or management of real estate, including disposal and acquisitions.

Cost is usually reported with reference to the building size but may also be expressed in terms of the number of desks or employees. As with capital expenditure, care must be taken in compiling operational costs, especially if being compared to other buildings or benchmarking.

Notes

1 www.gov.uk/government/statistical-data-sets/live-tables-on-energy-performance-of-buildings-certificates#epcs-for-non-domestic-properties
2 www.statista.com/topics/10049/construction-costs-in-the-uk/#topicOverview
3 www.lsh.co.uk/total-office-cost-survey

8   How to conduct a POE – people performance

In addition to subjective occupant feedback, there are objective measures of human performance that may be used in a POE to help better understand how the building supports the occupants. While valuable, these methods are more complicated and involve a considerable determined effort to achieve, so they are more appropriate for research than for a standard POE. However, these potentially useful methods are listed here for completeness.

Organisational performance measurement

When evaluating the impact of a building on its occupants, where practical, more objective measures could be used in addition to subjective occupant feedback. Performance measurement is the process of collating and reporting information to gauge the success and performance of an individual, group or organisation. As part of a POE, the link between building design and building user performance might be investigated.

Organisational, or business, metrics tend to be related to human resources, health and safety (H&S) or finance. Unfortunately, such metrics are confounded by factors other than building design, so they are rarely used as part of a POE. Caution is required if these metrics are to be included and ideally a control group, or benchmark, should be used as a comparison. Having a control group is a social research method when the response of a group affected by, in this case, a building project is compared with a similar group that is not part of the project. Alternatively, internal or external benchmarks could be used; for example, staff attrition rates and absenteeism are published by organisations such as the Chartered Institute of Personnel and Development (CIPD). Such metrics are also quite sensitive or commercial and confidential, so they are usually not readily available for including in a POE.

Researchers who use organisational performance measurements in their studies usually monetise any percentage increase in performance so that it can be included in a cost-benefit analysis. For example, in a workplace, the percentage increase in performance would be multiplied by the number of workers affected and then by their salary (or their anticipated income to the organisation) to calculate the potential additional income. Such an approach is adopted by the Institute of Workplace Management (IWFM) in the *Return on Workplace Investment* (ROWI) Tool

DOI: 10.1201/9781003350798-8

(Oseland, Tucker and Wilson, 2023). Including monetised performance gains, and the resulting return on investment, in a POE would be a clear way of highlighting the success and value of the building project. Unfortunately, this is not an easy task due to the nature of the data required.

There are many ways to measure individual, team and organisational performance, each with its own advantages and disadvantages. The more common, and objective, organisational performance metrics, grouped by human resources, H&S and finance are listed below.

Human resource metrics

- **Absenteeism** – To calculate the absenteeism rate, take the number of unexcused absences over a given time, usually a month or a year, then divide it by the total days in that period and multiply by 100. The HR team usually tracks this so it will be readily available, but nonetheless it may also be considered too sensitive to include in a POE. Absenteeism results in downtime and lost productivity. Clearly, if the building results in higher absenteeism, then there is a serious problem with the building. Unfortunately, despite being a good objective measure, absenteeism is affected by extraneous factors beyond those of the building project, such as seasonal viruses, transport and security issues. However, it can be benchmarked against external industry norms or, preferably, with other parts of the organisation. I recall BT showed how one of its newly refurbished buildings reduced absenteeism compared with the national average, but when I got permission to use absenteeism as part of one of my own POEs I found it had worsened due to an increase in flu that year.
- **Attrition and attraction** – Also known as staff turnover (churn) and retention. These metrics are also highly affected by factors outside those of the building project, such as the success of the organisation and the market. Overcoming high attrition involves additional advertising, recruitment and training cost to the organisation.
- **Job appraisal** – This is a combination of a human resource and a business metric. It is not commonly used in POE, but most organisations conduct individual performance appraisals, including checking whether people are meeting objectives and targets (like sales) or measures of creativity and innovation, such as patents filed. Some organisations offer performance-related pay, a monetisation of performance and ready-made objective metrics, albeit based on a subjective manager assessment. Clearly, a host of factors affect job performance other than the building, so it is not a reliable metric for use in POE.

Of course, subjective staff feedback may also be used to collate information on job satisfaction, self-assessed productivity, morale and wellbeing. Many occupant feedback surveys include questions asking how the workplace impacts the respondent's performance and wellbeing. A single rating scale may be presented or, increasingly, a set of questions asked in which the responses are combined to form a "productivity index". Some research evidence supports a relationship between productivity and ratings of satisfaction, and it is certainly more convenient to measure satisfaction than attempt to objectively measure performance.

Health and safety metrics

- **Absenteeism** – Absenteeism is included above, but it is also a key H&S metric, particularly when absenteeism is linked to an accident or incident.
- **Accident rate** – Some organisations monitor fatal accidents, reported non-fatal accidents and "near misses" daily. As well as accident rates, injury rates and frequency rates may also be reported. These various accident rates are also benchmarked, but they are calculated slightly differently depending on the country and industry sector. For example, the UK's Health and Safety Executive (HSE) publishes injury rates based on the number of people injured over a year in a group of 100,000 employees, and reports the frequency rate, the number of people injured over a year for each million hours worked by a group of employees.
- **Incident rate** – The incident rate or the number of days since the last incident or the loss of productive time (down time) are also key measures for some organisations. Note that an incident could be a malfunction of equipment (equipment breakdowns) or other failures rather than an accident *per se*. The productive days, the opposite of lost time, may also be reported.
- **Overtime** – The level of (paid and unpaid) overtime is an indication of the resources available and the stress placed on the staff and the organisation to complete the workload. Clearly, longer hours may lead to fatigue and potentially more accidents. Overtime is mentioned here for completeness, but it is not a common POE metric.

There are more H&S metrics and KPIs, but those mentioned above are the most relevant to building performance evaluation.

Finance metrics

- **Revenue** – The revenue, income or turnover of a corporate business, such as sales, is a common in-house metric of success. Again, revenue is affected by confounding factors, such as the quality of products and services, and it is also affected by the overheads of the business. Revenue (and profit) may be more relevant to organisations where the link between the building, facilities and equipment with output is less ambiguous, such as in manufacturing.
- **Profit** – Using profit or margin is similar to using revenue in a POE, but the overheads are accounted for.
- **Staff utilisation** – This is more commonly used by organisations with fee earners, such as law firms. The time spent on projects, so fee earning, may be expressed as an average and the ratio of fee earners to non-fee earners may also be reported. For call centres, the average time the staff spend answering calls and the number of calls answered plus other measures are used. Where appropriate, such metrics have the advantage of quantifying more knowledge-based work. Overtime, voluntary and paid, may also be used as a metric of financial

success. Like the other financial metrics, staff utilisation is also affected by factors other than the building, so it is rarely used in POE.

- **Performance delivery** – Other measures of success used by an organisation include time for completion, meeting deadlines, delivering within budget, good accuracy with minimal errors and high quality. Most organisations will have a key measure that makes obvious sense to them and is common across their industry sector. For example, in the insurance sector the number of claims processed will be key, whereas in the public sector the enquiries completed, or backlog, may be more useful. Research and academic institutions might use the annual number of publications, citations or patents as their main performance measurement. However, these are all highly affected by broader factors and are difficult to use in a POE.
- **Customer feedback** – New or repeat business is another measure of success, and customer satisfaction, recommendations and press coverage are also valid measures. For some organisations, the number of customer complaints will be a key metric. Customer feedback is quantifiable but more subjective than the above metrics and also subject to factors outside of the building project.

Like the human resource and H&S metrics, there are many more finance ones. Such metrics are more useful in detailed, long-term POEs and research projects than in a standard POE. This is partly because of their sensitivity or commerciality and partly due to confounding factors. Personally, I have had less success than others in linking the impact of the building design on these metrics, even in longitudinal studies and when using control groups. Nevertheless, I still believe they are metrics worthy of collation and reporting in a POE if readily available. It can be as useful to show there are zero significant effects on such key metrics as it is to show a positive or negative effect.

Cognitive and physiological measurements

Questionnaires can be used in POE to ask the respondents how the building, primarily their workplace, has affected their performance. However, to capture the information in a less subjective manner, some researchers are using cognitive performance tasks. In addition, some researchers have explored using physiological measurements to show the impact of buildings on people. Whilst previously such measurements were difficult and impractical, advances in wearable technology may make such data more accessible.

Cognitive performance measurement

The advent of online or app-based cognitive performance tasks has made the measurement more practical for use on a larger group of building occupants, and in situ. However, while such methods have been used as part of a POE, they tend to be only used as part of a broader study. For example, Gillen (2015) deployed several cognitive performance tests on office workers before and after they moved to National Grid's new workplace.

Cognitive performance refers to human cognitive abilities or mental processes, and it is a measure of how well our brain functions. It includes conducting the mental processes of perception, learning, memory, understanding, awareness, reasoning, judgement, intuition and language. Cognitive performance tasks mostly involve measuring the speed and accuracy of various tests of capability, which the Institute of Medicine (2015) categorises as:

- **cognitive and intellectual** – the capability for reasoning and problem-solving,
- **language and communication** – encompassing the ability to understand spoken or written language, communicate thoughts and follow directions,
- **learning and memory** – such as being able to register and store new information and retrieve information as needed,
- **attention and vigilance** – referring to the capacity to sustain focus and attention when distracted,
- **processing speed** – for example, the amount of time it takes to respond to questions and process information,
- **executive functioning** – an overarching term covering complex processes such as planning, prioritising, organising, decision making, task switching, error correction and mental flexibility,

Traditionally, cognitive performance tests were paper-based, but many are now available online or through apps. Examples of the more common tests used by researchers and practicing psychologists are described below.

- **Digit (span) memory test** – This is a popular test used to assess short-term working memory, which is the small amount of information that can be contained and used to perform further cognitive tasks. The test presents a sequence of numbers on the screen and the participant then attempts to recall them (Figure 8.1). If all the numbers are remembered correctly then the next sequence is one digit longer and so on.
- **Mental speed test** – A test of how quickly information can be processed and decisions made based on that information. The test consists of word/image pairs and simple mathematical equations where the participant notes whether they are correct or incorrect (Figure 8.1). The test complexity is increased by also occasionally asking for the answer to be reversed or "opposite".
- **Divergent thinking task** – Divergent thinking is the mental process of generating multiple creative ideas to solve a problem. Psychologist J.P. Guilford (1947) first coined the term "divergent thinking" after studying the ingenuity of US air force pilots during warfare. Lateral thinking is similar but involves an indirect and creative approach to problem-solving that is less obvious. There is a wide range of divergent thinking tasks available, mostly used to assess everyday creativity. Divergent thinking tasks ask participants to produce multiple ideas in response to pictorial or verbal stimuli. (My favourite is asking for uses for a house brick or a paperclip.) Points are awarded for the number of ideas and most original ones.

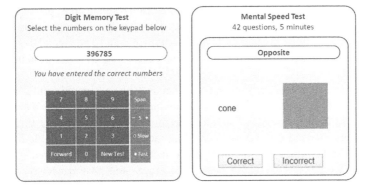

Figure 8.1 Examples or digit memory and mental speed tests

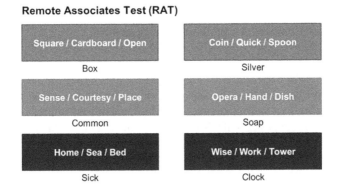

Figure 8.2 Example questions in the RAT

- **Remote associates test (RAT)** – This is another popular test to determine creative potential (Figure 8.2). The test typically consists of up to forty questions each of which consists of three common stimulus words that appear to be unrelated. The participant is then asked to think of a fourth word that is related to each of the first three. The associations gradually become more difficult, and the scores are calculated based on the number of correct questions.
- **Stroop test** – Developed in 1929, the original colour-and-word test presented a sheet of paper with a list of the names of colours printed in colours that do not match the spelled-out colour (Stroop, 1935). Now, the words are presented in coloured text on a screen. The test is used to measure an individual's cognitive processing speed, attention capacity and cognitive control. The contrast between the named colour and its conflicting coloured text increases cognitive load and the corresponding response time and accuracy.

- **Brain training apps** – There are numerous brain training apps, primarily developed to improve mental ability but they may have some application in POE. Apps such as those created by Peak,[1] Luminosity[2] and The Great Brain Experiment[3] use a mixture of exercises to test and score cognitive performance.

The choice of cognitive performance tests needs to be considered carefully. Firstly, they need to be relevant; for example, if studying an accountancy firm's new offices perhaps use digit memory and mental speed tests. Secondly, ensure the tests are easily accessible and that the data can be extracted in a format suitable for further analysis. Furthermore, due to confounding factors such as job-related stress, it is necessary to measure cognitive performance pre- and post-project on the same group of people and ideally compare the results with a control group.

The impact of the environment on creativity, innovation and cognitive performance is quite well documented, especially the benefits of natural environments and biophilic design. Demonstrating a link between a new workplace and cognitive performance is, therefore, a worthy pursuit, but alas like many of the other methods in this chapter it is not so straightforward to accomplish.

Physiological measurements

The health of the workforce is becoming more important – some organisations offer programmes to their employees to help maintain and improve their health. Ultimately this reduces absenteeism due to sickness and, for those organisations paying it, reduces private health care premiums. As part of a corporate health programme, employees may be requested to wear devices that monitor their health.

In my early days as a physiological measurement technician, I used ambulatory monitoring equipment to record the electrocardiogram (ECG), the heart's electrical activity and rhythm, of people undergoing psychotherapy. While technically mobile, the equipment was cumbersome and unreliable using wires, batteries and bulky recording devices, see example in Figure 8.3. The advent of wearable technology now makes the recording of health variables more accessible and practical.

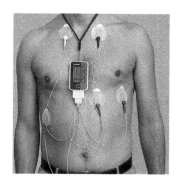

Figure 8.3 Ambulatory monitoring of ECG (Misscurry, Wikimedia Commons)

Physiological measuring instruments tend to record biosignals: a physiological pulse that can be continuously measured and monitored. Biometrics and biosignals refer to both electrical and non-electrical signals. Bioelectrical signals include electrocardiogram (ECG), electroencephalogram (EEG), electromyogram (EMG) and galvanic skin response (GSR), whereas non-electrical biometrics include heart rate (pulse), blood pressure, respiration (breathing) and body temperature. Interestingly, the polygraph, more commonly known as the lie detector, recorded several physiological measurements including blood pressure, pulse, respiration and GSR.

Modern wearable health devices are mostly worn on the wrist in the form of a smartwatch, like the Fitbit, Garmin and Apple Watch, but other devices are available, including glasses, rings, footwear, implants and technology embedded in clothing, see example in Figure 8.4. These small devices monitor biometrics such as heart rate, heart rhythm, blood oxygen, skin temperature, respiration and GSR as well as distance, exercise, calories burned and sleep patterns.

Some devices use contacts to sense electrical biosignals and skin conductance, whereas others feature optical detectors, which use LEDs to measure the reflectance of light from the skin. There are pros and cons for each mechanism, and they are better suited for different biometrics.

While wearable technology means that it is easier to monitor physiological variables across a group of people, in practice it is not so easy. Firstly, extracting the data from such devices is not straightforward because tech companies have become more protective of the data they collect. This means it is less likely that the data can be downloaded in a format ready for analysis, and therefore the participants are required to capture and pass on their data. Secondly, some physiological data may detect underlying health issues, and there is the whole ethical conundrum of how such information is handled and shared with participants, especially in academic research studies. Finally, like other techniques in this chapter, biometrics are confounded by many factors beyond the building design, and experimental methods such as pre- and post-project measurement and/or control groups would need to be deployed to make any sense of the data.

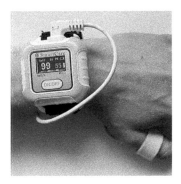

Figure 8.4 Early wearable pulse oximetry monitor (Charlton, Wikimedia Commons)

Notes

1   www.peak.net/
2   www.lumosity.com/en/
3   www.thegreatbrainexperiment.com/

9 How to report POE results

For a POE to be useful, it is necessary to analyse, interpret and present the data in an appropriate and meaningful way. The analysis needs to be at the appropriate level to provide credibility and rigour without unnecessary over-complexity. Similarly, the data needs to be presented in a proficient manner: it should be relevant and conveyed in a way that is easily digested.

Communication and balance

Several researchers have identified that people communicate using different styles, and it is necessary to appeal to those styles to ensure the information is well received. For example, Murphy (2015) categorised communication styles as: analytical – people who like to see numbers and hard data; intuitive – people who like the big picture and less detail; functional – people who like process and detail; and personal – people who like emotional language and personal connection (Figure 9.1).

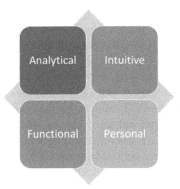

Figure 9.1 Four communication styles (Murphy, 2015)

DOI: 10.1201/9781003350798-9

It is, therefore, essential to consider the audience that will be reading the POE report or having it presented to them, and check whether they are the managing director or board-level officers, or from the corporate real estate, facilities management, HR or IT teams with potentially different communication styles. The people conducting the POE tend to be analytical or functional communicators, but they should reflect on the effect of the building on the lives of the occupants and the broader impact on the organisation, as well as the data and process, when presenting the results.

Fundamentally, the analysis and interpretation need to fulfil the POE objectives: identifying the project successes, any areas of concern, lessons learned and recommendations. All too often a POE that is viewed as a critique but a balanced evaluation, highlighting positive as well as any negative findings, will be better received and hopefully acted upon. Psychologists have long known that people recall more negative events than positive ones, possibly as part of an evolutionary process to help protect us (Cherry, 2022). To prevent a POE from being perceived as overly negative, it is therefore necessary to present as much, if not more, positive feedback than negative.

As well as analysing and interpreting the results, recommendations will also be expected. Recommendations might be grouped by short-, medium- and long-term reflecting resources and identifying any "quick wins", to show the respondents they have been listened to. The recommendations may be categorised according to different elements of the building design or broken down by the different types of occupants or their locations. Quite often, the recommendation may relate to how the building is used and the corresponding occupant behaviour, rather than to the physical design and conditions. For example, poor acoustics may be due to insufficient absorption, too low desk screens or lack of meeting spaces, but it may also be due to occupants holding meetings at their desks rather than using the meeting spaces provided, or due to them walking around the office speaking loudly on a mobile telephone, or simply shouting across the desk to colleagues.

The approach to the POE and the techniques adopted will also need to be explained. Key dates should be documented, particularly how long after occupation the POE was carried out. If a feedback survey was administered then the sample should be described, as should the decisions on which, if any, occupants were consulted through interviews and focus groups. Most technical POE techniques will also need to be fully described, such as the measurement locations of indoor environmental monitoring sensors and any measurement standards adhered to.

It is essential to share the POE results with the building users; for example, to show that their input is valued and that their time spent responding to a feedback survey has not been wasted. However, the information passed back to the occupants may need to be a slightly sanitised version of the main report, so that it is easy for everyone to understand, with any overly-sensitive issues omitted to help manage expectations. Nonetheless, negative findings should not be omitted or ignored but rather identified and the proposed actions explained, even if resolving such issues may require a longer-term solution and a short-term fix is not achievable for cost or practical reasons.

Graphical representation

POEs tend to produce vast amounts of data reported back in the form of tables, and the people who conduct detailed POE studies for research purposes often accompany the tables with statistical analyses (discussed at the end of this chapter). That is fine for a more academic audience and "analytical" communicators but may be overwhelming for others. In contrast, focusing on data directly relevant to the POE objective accompanied by a selection of clearly formatted charts will improve the report and presentation to the wider audience, making it more user-friendly, comprehensible and memorable.

Averages and totals

Quite often the data obtained from a POE will be presented as an average (mean, median or mode) or total (sum); for example, the average rating on a response scale in a feedback survey, the average office temperature, or the total energy or water used over a fixed period of time.

For clarity, the average may be illustrated graphically as a (horizontal) bar chart, see Figure 9.2, or the total may be presented with, for example, a column (vertical bar) chart, as used for Figure 7.8 in Chapter 7. A bar chart works better when there are lots of variables to compare or when the variable labels are lengthy because there is more space for the labels.

A drawback of a simple bar chart is that the average does not express the range or spread of the data, whether questionnaire responses or physical measurements; therefore, it is also important to present the standard deviation or quartile range. A boxplot, sometimes called a box and whisker chart, is an established way of illustrating the mean (white line), standard deviation (box) and range (whisker/line), as illustrated in Figure 9.3. However, these plots are quite confusing, so they tend to be used more in research papers than POE reports.

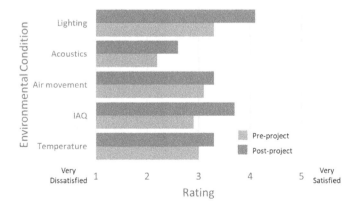

Figure 9.2 Clustered bar chart showing mean pre- and post-project survey responses

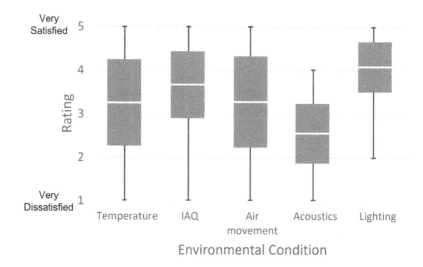

Figure 9.3 Boxplot of responses on a 5-point rating scale

When comparing pre- and post-project data the results of several key factors could be illustrated using a clustered bar chart, as in Figure 9.2. The second bar might also be used to compare the data with relevant benchmarks. Simple line graphs, favoured by experimental psychologists, can also be used to display pre- and post-project data or show the change of a key variable over several years. However, in occupant feedback research, line graphs are commonly used to show the difference in pre- and post-project between two or more groups, as illustrated in Figure 9.4. Line graphs are particularly useful for comparing the average responses of the people involved in the building project with those who were not, termed a control group in research studies. "Whiskers" might be added to the ends of the lines to indicate the range of responses.

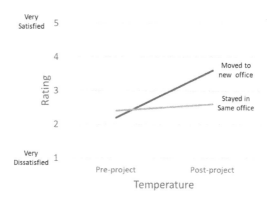

Figure 9.4 Simple line graph showing results of two groups of survey respondents

Caution is required when using averages to present questionnaire responses because it is not always clear what the response actually means. For example, on a 7-point rating scale, an average of, say, 4.6 means little other than it is above the mid-point and below the upper end-point. However, if the scale has each point labelled (for example, "1. very dissatisfied", "2. dissatisfied", "3. indifferent", "4. satisfied" and "5. very satisfied"), then it would be possible to report that an average of 4.6 lies between "satisfied" and "very satisfied".

This also illustrates why absolute values on different numbered scales cannot be directly compared. For instance, an average rating of 4 on a 7-point scale is like a rating of 3 on a 5-point one. Some researchers convert ratings on different scales so that they are comparable; for example, the rating on a 7-point scale would be multiplied by 71% to convert to a 5-point scale. Likewise, reporting averages can be confusing when using a mixture of unipolar and bipolar scales where the most positive rating may be at the end- or mid-point, respectively.

Furthermore, averages are more meaningful when the responses to a rating scale show a normal distribution, or classic bell curve – where the distribution of ratings is symmetric around the mean with more responses being a peak at the mean. It is meaningless to present the mean average of a bimodal distribution of responses with two peaks, as illustrated in Figure 9.5, because the mid-point does not represent the majority of responses. Either the modal average should be used, or possibly the respondents split into two groups to represent the two peaks of responses.

Frequency data

When analysing occupant surveys, an alternative to reporting the average is to report the frequency. For questionnaire responses, it makes much more sense to me to report the percentage of occupants satisfied rather than the average level of satisfaction. The percentage of responses to a question made on each rating scale point may be illustrated using a stacked bar chart, as shown in Figure 9.6. Such a chart works better when the same rating scale is used across a section of questions. Note that simple bar charts and pie charts can be used to display the range of answers to single questions, as well as to illustrate background variables such as gender or age groups.

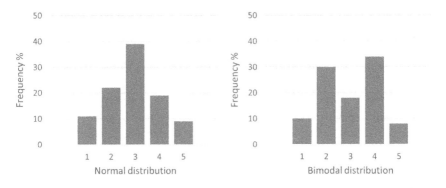

Figure 9.5 Normal and bimodal distribution of responses to a 5-point rating scale

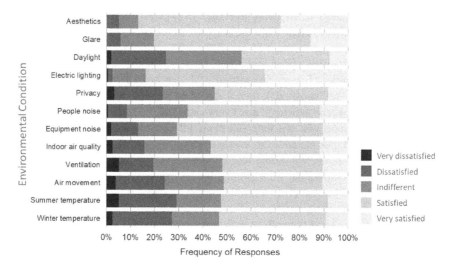

Figure 9.6 Stacked bar chart of survey responses

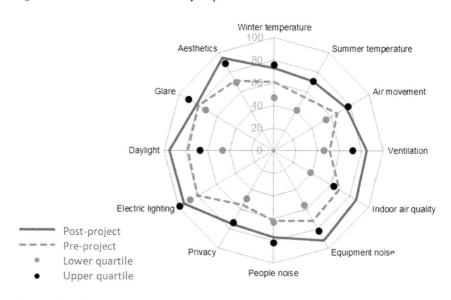

Figure 9.7 Radial chart of pre- and post-project survey responses on a scale of 0–100% satisfied

While useful, it is evident that a chart such as Figure 9.6 portrays lots of information, making it somewhat difficult to interpret. One option is to reduce the responses on 5-point (or more) rating scales to three or even a single response and display that. For example, a single "percentage satisfied" rating could be produced by combining the percentage frequency of ratings of "satisfied" and "very satisfied" and, for ease of display, temporarily disregarding the other ratings. The results could be presented as a bar, line or possibly a radial chart, known as a "spider graph" (illustrated in Figure 9.7), which provides a "fingerprint" of the building. Like bar charts, there are different styles of radial charts. Because the frequency data has been reduced to a single response, the radial chart makes it simpler to plot multiple sets of responses and make comparisons, such as between pre- and post-project responses, different groups of respondents or against a benchmark. Averages can also be presented on radial charts.

GIS mapping

GIS can be blended with maps to provide a visual representation of the data using colour, referred to as heatmaps. Traditionally, GIS was used at the urban level, but some researchers use it to map data onto building plans. For example, the plan could be used to show building utilisation, as in Figure 9.8 (often coloured blue to red), temperature or lighting measurements, or even overall occupant survey responses. This approach is particularly good for identifying issues located in certain areas of a building.

Analysing open-ended questions

The data collated and presented in a POE is often quantitative, such as physical measurement or survey ratings, but may also be qualitative, as with the response to open-ended questions or comments made in surveys or interviews. The qualitative data may be used to capture specific comments, especially from key stakeholders, that optimise or highlight the quantitative findings. However, qualitative data can also be quantified.

Qualitative data is often subjected to some form of content analysis, where similar responses are coded and categorised to create themes, as proposed by Saldaña (2015) and shown in Figure 9.9. For example, "my desk is too small", "the desk is the wrong shape" and "my desk gives me backache" might be grouped together and called "desk ergonomics". The categories and themes may be counted and reported back as tabulated frequencies. The frequency of similar responses can then be plotted on a bar or pie chart. In addition to content analysis, qualitative researchers may use related methods such as narrative, thematic, discourse, framework, grounded theory and interpretive phenomenological analysis.

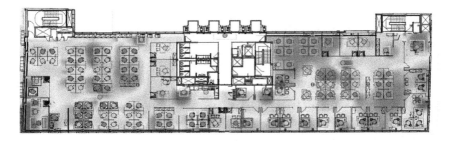

Figure 9.8 Heatmap of building utilisation (where the darker shading represents "hotspots" of activity)

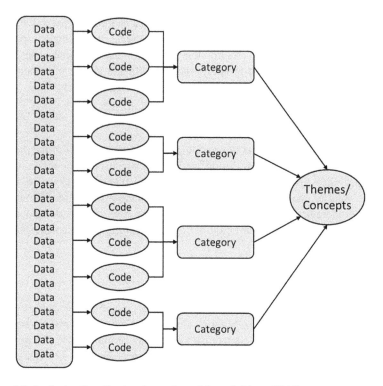

Figure 9.9 Analysis of qualitative data, adapted from Saldaña (2015)

Such analysis may be conducted by hand, but software is available to explore and interpret large data sets. While qualitative data is rich in content, manual analysis is time-consuming, so it is advisable to limit the number of open-ended questions when surveying large groups unless familiar with the recognised software tools.

Recent developments in natural language processing and machine learning technologies mean that AI platforms are starting to be used to interrogate, structure and interpret qualitative data at speed and scale. For example, as mentioned in Chapter 1, Audiem has created a platform to analyse and interpret large quantities of free-text responses, including open-ended questions, help desk comments and chat room threads.

Comparisons and benchmarking

If the POE is being used to assess the impact of a building project on the occupants, then ideally the various forms of data, such as feedback surveys, will have been conducted pre-project and post-project allowing a direct comparison. Similarly, predicted performance modelled prior to, or in the early stages of, a project would be compared with the data recorded after project completion once the building has been in use for some time.

Unfortunately, this is often not possible, and if the POE was conducted post-project only, then it may not be clear whether the data indicates improvement or not because the starting point is unknown. In a feedback survey, for example, finding 85% satisfaction with a particular aspect of the building may seem high, but there is no guarantee it was not higher before. Likewise, not conducting a sustainability audit prior to a refurbishment, or not modelling the performance ahead of the project, means there is no data to compare against.

Benchmarking

If there is no pre-project data, the alternative is to benchmark the result with those found in the rest of the occupier's property portfolio, or against the buildings of other similar organisations. Consultants using tried and tested POE methodologies will have access to existing benchmark data. Note, however, that benchmarks are more valid and reliable when comparing building performance within the same property portfolio because there is some consistency in the nonphysical factors, such as building management and maintenance, or organisational structure and culture. The former may have more impact on sustainability and environmental measurements and the latter on occupant feedback.

Whether using an existing database for benchmarking or creating a new one, the same methodology must be used. It is extremely difficult to compare the results from different questionnaires and different measurement techniques. In feedback surveys, even a slight change in the wording of a question can change the meaning of that question, making it difficult to compare it with other similar ones. Furthermore, it is extremely difficult to compare the ratings of building performance made on different types of scales. Likewise, environmental conditions and other compiled data need to use consistent measurements to be comparable.

As mentioned, POE questionnaire data is usually based on an average rating or a frequency score. To create a benchmark, the data from each individual questionnaire is entered into a database, and the benchmark for each question can then be derived from the individual responses across all buildings. Alternatively, the benchmark could be derived from the collective responses grouped for each building so that a building score is created for each question. The same approach could be applied to data collated using other POE techniques. The benchmark average rating or frequency score is usually accompanied by an indicator of the range of responses, such as standard deviation or quartile ranges.

As with other data, the benchmarks may be shown graphically. For example, the key in Figure 9.7 explains that the grey dots represent the upper quartile of the feedback survey database, and the black dots are the lower quartile. A lower quartile dot represents the highest score of the worst-performing 25% of buildings, and the upper quartile is the lowest score of the best-performing 25%.

For a more relevant comparison, once a high number of POEs have been conducted and captured in a database it may be possible to develop a series of benchmarks based on sub-sets related to the different characteristics of the buildings and organisations surveyed. For example, benchmarks might be created for building features, including the building services such as naturally ventilated versus air-conditioned; for tenure such as owned versus leased; for the development such as bespoke versus speculative; or perhaps based on geographical location. Similarly, benchmarks based on the organisation may include the type of business sector, such as technology versus finance, or public versus corporate etc.

Another useful way of benchmarking the questionnaire results is to order all the buildings in the database according to their overall rating of satisfaction, or another index, then highlight the building being assessed, shown in Figure 9.10. This process could be repeated for different groups of buildings, such as particular business sectors.

In some circumstances, it is useful to compare the occupant feedback, such as satisfaction ratings, with the other types of data collected as part of the POE (such as cost or space metrics). For example, Figure 9.11 shows the percentage of overall satisfaction plotted against the fit-out cost for the first twelve (refurbished or newly fitted-out) offices that I evaluated. In simple terms, value is a ratio of quality to cost, or $V = Q \div £$, representing a return on investment. Comparing metrics of quality, such as occupant satisfaction, self-assessed productivity or expert ratings, against cost metrics, whether the project or total occupancy cost, is one means of illustrating the value of the building's performance. In Figure 9.11, the higher-value building projects are represented by the lighter dots whereas the darker dot represents a poor-value project.

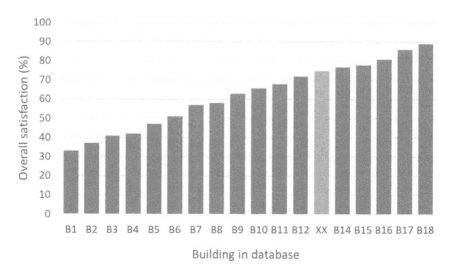

Figure 9.10 Buildings in a database ordered by overall satisfaction

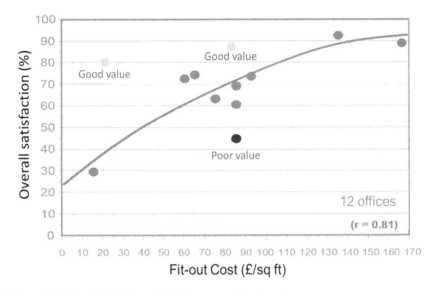

Figure 9.11 Overall building satisfaction versus fit-out cost

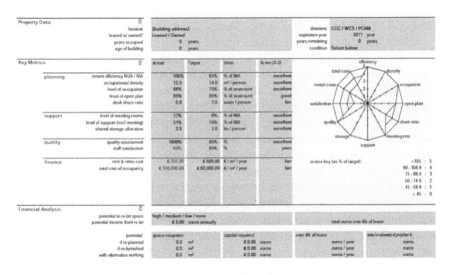

Figure 9.12 Example of building evaluation dashboard

Combined metrics

Some organisations provide dashboards that display the key metrics and measurements obtained through an ongoing POE into one convenient spreadsheet, like the one shown in Figure 9.12. This is an opportunity to compile and compare subjective metrics such as satisfaction and quality with more objective ones such as cost and space. An additional step is to set targets and show how the individual metrics for each building perform against those targets. Such dashboards are particularly useful for organisations with a portfolio, or campus, of buildings, and highlight the performance and benefits of a building beyond cost and space reduction.

Statistical analysis

Statistical analyses can be broadly divided into descriptive and inferential statistics. Some descriptive statistics have already been mentioned, including the average and frequency of survey responses along with the standard deviation and quartile range. Descriptive statistics are basically used to describe, organise and summarise the data.

The rest of this section relates to inferential statistics, which are used to draw conclusions and make predictions based on the collated data. When the data is based on a sample, such as a feedback survey or spot measurements, then inferential statistics are used to generalise the results to the larger population (people or other data) represented by that sample. Those statistical tests that corroborate the differences between groups of data or substantiate correlations (relationships) between variables, along with quoting probabilities or significance levels, are inferential statistics.

Statistical significance

Statistical significance represents the likelihood that the survey results did not occur by chance and that there is high confidence the result is real rather than an error. Statisticians aim to avoid Type I and Type II errors, where Type I is a false-positive or accepting a result that may have occurred by chance, and Type II is a false-negative or rejecting a result that is likely to be true. Furthermore, statistical significance determines whether, and to what extent, the results from a sample really represent those of the broader target population rather than happening to match them by chance. Most social scientists assume that a result is representative when there is a probability (p) of less than 5% that the result is due to chance, usually shown as "$p<0.05$".

When conducting feedback surveys, ideally all the building occupants should be invited to participate. If all those invited actually responded, then the sample is the same as the building population. As such, there would be less need to calculate whether the responses to particular questions are statistically significant. However, it is very rare that all of those invited to participate in a survey do respond, and so the data collected will, in effect, be a sample. Furthermore, POEs often involves testing the difference in response between various groups of occupants, which are also samples.

The larger the sample, the more likely the test will yield statistically significant results, providing higher confidence that the result did not occur by chance.

Size of effect

The term "statistically significant" is often misunderstood. Although it is a measure of the accuracy of the data, it does not necessarily mean that the strength of the relationship or size of the difference between two variables is sufficiently large to be considered relevant or important. For example, in large samples the correlation between two variables may be found to be statistically significant even though the relationship is weak. The strength of the relationship is referred to as "effect size", or "practical significance", and there are separate statistical tests for effect size such as Cohen's d and Pearson's r tests.

When testing the difference between two groups, for example when using analysis of variance, Cohen's d is used. The rule of thumb interpretation is that a d value of 0.2 indicates a small effect, 0.5 and above a medium effect and 0.8 a large effect. For relationships between two variables, a Pearson's r of 0.1 indicates a small effect, 0.3 and above a medium effect and 0.5 a large effect.

Choosing statistical tests

Analysis of POE techniques, especially questionnaire data, is a complex task requiring a high level of expertise in statistics. This section summarises the main statistical tests used for analysing POE data, particularly questionnaires.

In order to predict the parameters (such as responses) of the full population from a sample, it is necessary to know certain factors about that population. Inferential statistics that depend upon knowing, or more often assuming, certain population parameters are called parametric tests. To use parametric tests, the data should meet the following criteria.

- **Normality** – normally distributed (bell curve) with minimal skewness (a measure of lack of symmetry), kurtosis (a measure of how pointy or flat the distribution is) and no outliers.
- **Equal variance** – exhibit similar degrees of homogeneity of variance.
- **Independence** – randomly and independently sampled from the population.

In theory, descriptive tests should be conducted to test that the sample characteristics match those of the population before proceeding with parametric tests. Nonparametric tests also exist and do not require such stringent assumptions of the population parameters. Non-parametric tests may therefore be considered safer to use than parametric tests.

The type of data collated also determines the type of statistical test required to analyse it. The numbers assigned to a measurement, whether a questionnaire response or the reading from a technical instrument, are classified into four distinct levels, referred to as scales of measurement, shown below. Each level has restrictions on how it may be analysed – the two higher levels of measurement scale may be analysed using parametric statistics, whereas the two lower levels are restricted to analysis using non-parametric statistics. Note that each level is also hierarchical; that is each higher level incorporates the properties of the level below it.

- **Nominal scale** – This is when numbers are used to code categories of information; for example, assigning 1 for "air conditioning" and 2 for "natural ventilation". The standard "Yes/No" scale is a dichotomous nominal scale. The coding does not indicate that one is better than the other and as such the numbers have no numerical meaning so they cannot be added or subtracted and so on. The numbers must be mutually exclusive, so represent only one category, and be exhaustive, so include all possible categories.

- **Ordinal scale** – This scale is used to code categories, as above, but also to indicate the rank order of each of the items. For example, buildings may be coded as 1 for "small floorplate", 2 for "medium floorplate" and 3 for "large floorplate". The numbers do not indicate absolute quantities and it cannot be assumed that 2 on an ordinal scale is twice 1 or that 3 is three times the value of 1. Nor can it be assumed that the intervals between the three numbers are equal; in this example, the "large floorplate" could be four times the size of a medium floorplate, whereas a medium floorplate may only be twice the size of a small floorplate. Ranked responses are ordinal data; likewise, semantic differential and labelled Likert-type scales are technically ordinal scales, even though many researchers treat them as interval.

- **Interval scale** – In addition to reflecting an order to the data, the interval between each assigned number is of equal quantity. The Celsius and Fahrenheit temperature scales are classic examples of interval numbers. Although the intervals are equal, zero degrees is arbitrary as it does not mean there is no temperature *per se*; in contrast, the Kelvin scale is ratio data as zero represents absolute zero. Indices calculated from ordinal scales are also usually considered to be interval data. Interval scale numbers can be added or subtracted, but as there is no true zero they cannot be meaningfully multiplied or divided. Thus, the interval between 20 and 25°C is the same as that between 25 and 30°C, but 30°C cannot be considered twice as hot as 15°C. If it were, a conflict would be created because 15°C equals 59°F and 30°C equals 86°F, and clearly 86 is not twice 59.

- **Ratio scale** – These scales represent the highest level of measurement and have all the features of the nominal, ordinal and interval scales. In addition, ratio scales include a true zero and have meaningful ratios between arbitrary pairs of numbers. Therefore, all the arithmetic operations (addition, subtraction, multiplication and division) are meaningful. Rating scales do not represent the ratio level of measurement; rather it is more associated with physical measures such as space metrics. Ungrouped background variables such as age or time in the company or building are examples of ratio measures. Be careful of how logarithm scales, such as sound level measured in decibels (dB), are handled.

Some researchers, especially social scientists and advocates of semantic differential scales, consider rating scales to be interval rather than ordinal data which has the advantage of allowing parametric statistical tests to be used in their analysis. It is assumed that the respondent will interpret the scale points to be equidistant, and this is more likely if the scale is symmetrical with meaningful labels around a neutral point. However, academics continue to debate whether rating scales are ordinal or interval, with more purist researchers arguing that the intervals on these scales are not always treated as equal by the respondents and, therefore, parametric statistics should not be used. In practice, if the data meets the criteria for parametric tests, then there is more justification for using them, and some statisticians argue that parametric tests are less prone to errors than non-parametric ones. For precaution, I recommend initially conducting basic parametric and non-parametric tests and if they produce similar statistical results then move on to more complex parametric tests. Of course, this may all be academic if the survey sample is sufficient to fully represent the building population.

Figure 9.13 is a flow chart summarising the main statistical tests used for analysing questionnaires and other data. For example, the main descriptive statistics performed on ordinal data should be the median, mode and quartile range. In comparison, the mean and standard deviation can be used with interval data.

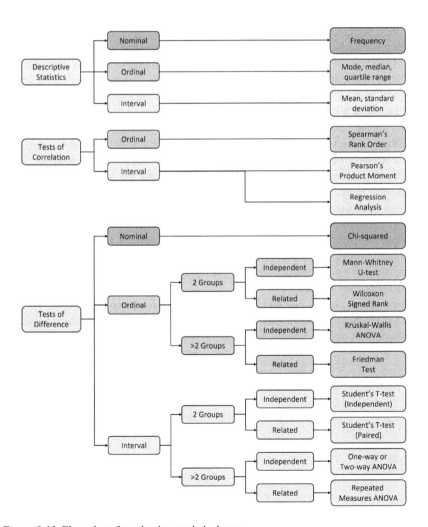

Figure 9.13 Flow chart for selecting statistical tests

All statistical tests can be performed on ratio data, but parametric tests are more suitable.

The inferential statistics used on ordinal data are non-parametric tests, including the Chi-squared test and Mann-Whitney U-test to verify differences between groups, and the Spearman's Rank Order for testing correlations. In contrast, interval data is associated with parametric tests of difference such as the Student's T-test and ANOVA, and tests of correlation, such as Pearson's Product Moment and Regression Analysis. The appropriate statistical test also depends on the number of groups being compared and whether the groups are independent of each other, for example air-conditioned versus naturally ventilated buildings, or related such as the same building evaluated over time.

Cleaning data

When a researcher receives new data, they are often keen to start analysing it to unearth potentially interesting results. However, my experience has demonstrated the importance of preparing, checking and "cleaning" data before commencing analysis. It is quite disappointing to discover a statistically significant result and then later find it is based on wrongly entered or miscoded data.

So, the data and its coding need to be thoroughly checked before analysis commences, especially questionnaire data. For example, check how "not applicable" responses have been coded, and ensure they are excluded from the analysis. Also, check that genuine missing data has been left blank and not been coded has a zero, as this will affect averages and so on. A simple first step is to conduct a frequency analysis of all the variables to see which codes are associated with them and if there are any anomalies.

More sophisticated statistics packages, like the Statistical Package for the Social Sciences (SPSS), will require the type of data, as listed in the previous section, to be assigned to each variable. This has the advantage of guiding the type of statistical tests that can be used.

Any indices should be calculated once the data is cleaned to save on recalculation if errors are spotted later. If several databases are to be combined, such as questionnaires and IEQ data, check the marrying up of the data sets.

10 Where to conduct a POE

Location and building type

The main emphasis of this book has been workplaces, particularly offices, and there are many POEs of such spaces, including the UK's National Audit Office's (Concerto Consulting, 2006) compendium of six corporate and public sector offices, the study of 22 US government office buildings by Fowler et al. (2011) and Park, Loftness and Aziz's (2018) review of 64 North American office buildings. But, of course, a POE can be conducted in any occupied building. I even have a vague memory of a POE being carried out in a pig pen, but it was most likely part of a broader research study on animal husbandry. The following are just a few examples of case studies of POEs on human-occupied spaces, other than offices, that are published worldwide:

- housing (Gupta and Gregg, 2020; Ozturk, Arayici and Coates, 2012; Palmer, Terry and Armitage, 2016; Sanni-Anibire, Hassanain and Al-Hammad, 2016; Stevenson, 2019; Turner, 2006),
- hospitals and health centres (Kalantari and Snell, 2017; McLaughlin, 1975; Thomazoni, Ornstein and Ono, 2016; Zuo, Yuan and Pullen, 2011),
- libraries (Enright, 2002; Hassanain and Mudhei, 2006; Lackney and Zajfen, 2005),
- nurseries/crèches (CABE, 2008),
- parliament, such as – House of Commons (Schoenefeldt, 2019),
- police stations (Home Office, 2009),
- prisons (Wener, 1994),
- research centres (Heerwagen and Zagreus, 2005),
- schools (Abdou and Al Dghaimat, 2016; Markus, 1972; Watson and Thomson, 2005) and see the New Zealand's Ministry of Education website for POEs of schools there,[1]
- shopping malls (Salami, Akande and Oke, 2022),
- universities (Hadi and Kiruthiga, 2008; Ikediashi, Udo and Ofoegbu, 2020) and see Nottingham University's website for POEs of its buildings.[2]

Innovate UK (Palmer, Terry and Armitage, 2016) conducted a BPE of 40 UK buildings, including community centres, hotels, libraries, offices, schools and

DOI: 10.1201/9781003350798-10

supermarkets. However, their focus was on sustainability variables rather than occupant feedback. In their *State of the Nation Review*, Gupta and Gregg (2020) conducted a meta-analysis of an impressive number of POEs of UK homes. The main focus was on actual versus predicted performance and sustainability, but a section on occupant feedback was also included.

There be many more unpublished case studies, some of which have been presented at conferences. The remainder of this chapter includes case studies, mostly focusing on occupant feedback in offices across the world. The case studies, contributed by the POE specialists who conducted the studies, highlight the different POE techniques and interpretations of POE.

Notes

1 www.education.govt.nz/school/property-and-transport/projects-and-design/design/designing-learning-environments/post-occupancy-evaluations-of-school-building-projects/
2 www.nottingham.ac.uk/estates/development/post-occupancy-evaluations.aspx

Case study A: Standard Chartered success story in Taiwan

Tim Oldman
Leesman, UK

Providing exceptional workplace experience has long been at the forefront of the property agenda of Standard Chartered in support of the bank's ambition to create an inclusive culture, enabling colleagues to innovate together, generate new ideas, solve problems and continuously improve.

As planning began for its next phase of workplace innovation, the Covid-19 pandemic hit. This presented an abundance of challenges for many organisations; however, for Standard Chartered, it helped shape the strategies for its future office design and workplace experience operations. As such, everything was redefined to one simple strategy: twice the experience in half the space. The strategy aims to maximise productivity as well as enhance user experience and derives value from the facilities that deliver both.

Standard Chartered adopted the Leesman approach for measuring employee workplace experience across its global office portfolio. Using insights from a Leesman Office survey of its portfolio in 2019 (such as low employee satisfaction scores for "relaxing/taking a break", "video conferencing" and "restaurant/canteen"), the Property team could focus on improving the elements of the office that could entice employees back to the office environment after the Covid-driven period of home working.

Workplace experience transformation in Taiwan

In December 2021, Standard Chartered consolidated three Taiwan offices into a brand-new, state-of-the-art headquarters. Having been occupied for 25 years, the buildings were outdated, with limited communal spaces, segregated departments, and individual personalised workstations. However, the pre-Covid survey results for the Taiwan buildings indicated that employees longed for a space that would promote collaboration and flow within the office. Shelley Boland, Regional Head of Property Asia, explains: "The survey highlighted a number of aspects which needed to be prioritised as part of the new headquarters, helping us to refine our strategy to focus on improved wellbeing facilities and enabling collaboration and interaction with colleagues".

It was about change; from a historically siloed way of working where every employee had an individual space. These insights were incorporated into the organisation's redefined global design guidelines. The objective of the design for the headquarters was to ensure that important elements pertaining to wellbeing, productivity, sustainability, accessibility and a sense of community would be curated within the new office space.

Wellbeing was an essential factor that the organisation wanted to address. According to Vikey Hogan, Head of Property in Taiwan, Standard Chartered wanted to provide a space that would allow seamless collaboration, better flow within the building and more integration within traditionally siloed departments. Moving into the new era of office design, it was crucial to provide a space where employees could feel better supported and more connected – to really bring people together.

Standard Chartered's success was verified by a Leesman Office survey conducted in May 2022. The survey showed major improvements in employees' Lmi scores, see Figure 10.1. According to the survey insights, there are seven key areas in the new office design that have improved the workplace experience: wellbeing, catering, technology and IT support, community hosts, change management, sustainability and accessibility.

A greater focus on wellbeing

Wellbeing was a key aspect in the design guidelines for the new headquarters, with Standard Chartered dedicating 3–5% of floor space to wellbeing services and features. Innovative features include rejuvenation areas, quiet zones and massage chairs where employees can unwind and an entertainment area with a basketball arcade, punch bags and dartboards, a dedicated parents' room, and a medical centre.

Natural light and greenery are some of the other elements that contribute to employee satisfaction with the new headquarters' wellness offering. In fact, it has been one of the biggest satisfaction drivers, which illustrates the biophilia hypotheses around humans' innate tendency to seek connections with nature.

Figure 10.1 Leesman Office Survey productivity results 2019 and 2022

Promoting community culture through culinary experience

A key area of focus was nourishment, with micro-markets aspiring to "vertically activate" the building and promote better flow and integration, as employees are encouraged to move up and down to get their favourite food. This has been extremely well received, see Figure 10.2, and proved to be hugely successful, especially when considering the many street food vendors in the area selling food at very competitive prices.

The micro-markets not only provide a good variety of healthy meal and snack options at competitive prices, but they are also convenient for those working non-standard hours. The food and drinks offered at the Work Café promote a healthy choice by meeting necessary health recommendations, focusing on minimum sugar and salt.

Enhanced technology and IT support

In an office with over 1,000 employees, technology and IT play a significant role in productivity. Standard Chartered innovated to create an enhanced IT helpdesk, the Guru Bar, where employees can walk into the space, speak with a person and solve any IT-related problems efficiently on the spot.

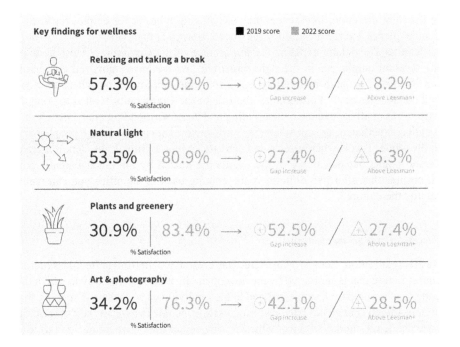

Figure 10.2 Leesman Office Survey wellness results 2019 and 2022

Jennifer Henderson, Global Head of Workplace Experience, summarises the value of this new addition: "With IT issues, you don't always know how to explain what's broken and what's not working, or ask for a specific thing that you need. So being able to go up to someone directly with your computer, who can help you and solve the issue right on the spot, has been another huge productivity enhancer. Instead of calling the helpline, employees can simply pay a visit to the guru bar on the destination floor and enjoy a coffee while waiting for their laptop to be fixed".

Day-to-day exceptional experience

With lockdowns causing employees to adjust to working from home, the need to connect and spend time with teams has grown significantly as employees have realised this is something they cannot do in their home offices. Jennifer Henderson explains: "People come back to collaborate, to spend time with their teams, and to generally have a really good experience".

Based on these needs, Community Hosts have been introduced and are one of the ways in which Standard Chartered plans to "earn the commute" – where their workplaces are worth the expense, effort and time put into getting there. "Because you get out of bed, you get properly dressed, you pay for your commute. So, something needs to add up for you to make that trip", says Jennifer Henderson.

Supporting the adjustment of removing individual offices and implementing a shared working environment with team zones, the Community Host is key in making employees feel supported by helping with any questions and pointing them in the right direction. In essence, they are making it easier for employees to find people, places and the technology they need to work at their best.

Jennifer Henderson explains the importance of the Community Host in helping to point employees in the right direction: "As we have rightsized the space, it's important to have community hosts to guide employees on where to work and collaborate best". Furthermore, the role of the Community Host is to connect employees to all the events in the building and in the surrounding area. Post-Covid, location experiences are considered a "pull" factor that can encourage people back to the office. This is where events and activations play a critical role. Jennifer Henderson points out that many of the events are hosted by the management team: "It ensures that after two difficult years, coming back to the office you can really rebuild the culture".

Successful change management

As an organisation, Standard Chartered has created a better-integrated, less hierarchal structure that is supported by the new open office design. The way in which the company has prepared, supported and helped individuals and teams in making this organisational change has been inspiring. Asking people to give up their offices can pose challenges and resistance, which is why senior management decided to lead by example by being the first to give up their offices.

The reduction of individual offices has helped Standard Chartered to eliminate excess layers of space management and has improved the speed and coordination of communication between employees. The new office has a state-of-the-art desk and room booking system, enabling flexibility and ease of reservation within the building, as well as a smart sensor system that will provide data insights to enable continued enhancements.

Jennifer Henderson explains: "It's about moving from 'me' to 'we' and getting people into the mindset of sharing. What helped tremendously was our Taiwan CEO leading by example. He was the first one who gave up his office. He sits out with his team, which he admits took a little while to get used to, but he absolutely loves it and says that it was one of the most positive changes he has made. It's fantastic because, at Standard Chartered (SCB) we are on a mission now to convert all personnel offices into shared spaces to encourage a much flatter hierarchy. We've made amazing progress globally".

Sustainability – top of the corporate agenda

Climate change has driven the awareness and expectations of sustainability to the top of the corporate agenda. For Standard Chartered, steps that have been implemented to promote sustainable spaces include the installation of electric vehicle charging stations in car parks, using more energy-efficient LED lights and eco-friendly furniture throughout the office design, implementing food composters for recycling, and having washroom basins fitted with water-efficient nozzles.

According to Vikey Hogan, the Taiwan HQ is already in an advanced stage of being evaluated, and the company is expecting to get a LEED certification in the fourth quarter of 2022: "It helps with our agenda of accelerating to net zero carbon here in Taiwan. It is a more carbon-efficient building than our previous building, and I think it raises awareness in that sense as well. Over the course of this year, we'll receive a LEED certification at an appropriate level".

Improved accessibility

As part of the wider Standard Chartered goal to become a disability-confident organisation, making the office accessible to everyone is paramount to an inclusive environment. It welcomes all colleagues to fully unlock their strengths by being part of an integrated community and improves workplace culture, which in turn enhances productivity, retention and motivation for employees.

Standard Chartered has incorporated accessibility features for disabled employees to ensure they have the tools and guidance they need to function and perform in the workplace. Part of this is an indoor way-finding system, as Vikey Hogan explains: "We have visually impaired colleagues in our contact centre; to cater for them, we have way finding aids in the whole building, including the floors, meeting rooms and lifts".

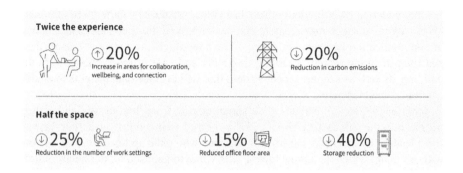

Figure 10.3 Leesman Office Survey overall results

Conclusion

The findings from Standard Chartered's most recent survey are a testament to the incredible transformation and impactful improvements it has made, not only in terms of the improved Lmi scores, but also through gaining a Leesman+ certification.

The office has scored better than the home across most business functions, while corporate pride has improved by 33.6%. This was the first standard measurement on a workplace that has implemented Standard Chartered's "twice the experience in half the space" strategy and has proven to be successful, with Lmi scores from three buildings increasing as employees have transitioned to a single headquarters.

Taking all the improvements into account, the new Taiwan headquarters is an attractive working environment for Standard Chartered employees compared with working from home, making the office a magnet, not a mandate.

Case study B: A large office building in the Netherlands

Susanne Colenberg
Delft University of Technology, Department of Human-Centered Design, Netherlands

Background

The Dutch government office was built in the early 1990s by a renowned architect and a quarter of a century later it needed a drastic update to new sustainability standards and working practices. The annexe was demolished, a large atrium was constructed, and new glass façades were installed. On 1 November 2017, the 80,000 m² office was officially opened by the King after three years of renovation, which was designed by another renowned architect.

In the four months before the official opening, two ministries and several administrative organisations had gradually moved in. For some employees, the concept of hot-desking was new, others were already used to it. However, for all of them, sharing the work environment with other organisations was a new experience.

The renovated office had to be energy efficient and accommodate 6,000 employees by offering 3,100 desks and 2,200 seats in lounge areas and meeting spaces. The main entrance of the building is at street level, tucked away in a passage from the train station to the city centre. Floors one through five feature general spaces, such as formal meeting rooms, the service desk and the restaurant. Floors 6–15 are office floors featuring a standardised mixture of open and enclosed workspaces, small meeting rooms and lounge areas. Despite the flexible office concept, each floor has been assigned to one of the organisations. The top floor featured large double-height open workspaces, and some indoor trees, and was intended to catch the overflow of other working floors.

Figure 10.4 Central entrance to the working floors – the fourth floor is a dedicated meeting area (Photo: Sensoy Kaan)

Soon after the opening, the project received an award for outstanding architecture and sustainability. Meanwhile, many users complained about crowding, cold and other discomforts. Their complaints brought forward the start of the POE, which was originally planned for six months to one year after moving in. The Ministry of Internal Affairs commissioned an evaluation from the user perspective by the independent Center for People and Buildings (CfPB), a competence centre on working environments where I was working at that time. We were granted a budget for a thorough evaluation because this office building was the first showcase of a new office space policy for civil servants and the Ministry wanted to know how it had worked out. Additionally, the POE had to provide input for a fact-based discussion with users and providers.

Approach

At CfPB, we assembled a research team of three experienced workplace researchers, research assistants and several experts on the subject. The research design featured the triangulation of different methods to obtain reliable and rich data.

First, the occupancy and use (computer work, conversations, phone calls, etc.) of workstations and meeting areas on every office floor were observed for two weeks, eight times a day. A workstation or seat was registered as occupied if a person was using it at that moment of observation or if it showed clear signs of usage (for example, if the computer display was on, a coat was on the chair or personal belongings were on the table). We counted the number of workstations and meeting seats per floor and registered their type (room capacity and open vs. enclosed) to check if the organisation's standards were being met.

Second, we constructed an online survey which included the standardised Work Environment Diagnosis Instrument, developed by CfPB, and additional questions about flexible working and specific services, as requested by the Ministry. Including the standardised questionnaire meant that we could benchmark the building against 72 flexible offices (that is organisations having desk-sharing policies) in the database of CfPB. An anonymous link to the survey was distributed by email to all 7,000 employees of the occupant organisations. After analysing the main issues arising from the survey, we conducted 27 interviews with groups of 3–8 employees representing a variety of departments to gather details about the experienced positive and negative aspects of the physical work environment and the users' own ideas for solutions. Managers were interviewed in separate groups to increase confidentiality.

Finally, in a team of three office usability experts, we rated the building on the criteria of the national standard for the usability of public buildings, which covers accessibility, flexibility, safety, comfort and facilities. This standard did not cover usability from the perspective of the individual user and did not include interior design. Additionally, we used a list of quality criteria phrased by the Ministry to guide output specifications. This list featured 50 relatively easy-to-apply features that were supposed to increase usability. We based our ratings on available documentation and a visit to the building where we used apps on our telephone to get an indication of light levels. All in all, it took us five months to plan the research, collect and analyse the data, and write the 75-page report.

Results

The observations showed that on Mondays, Tuesdays and Thursdays, the occupancy rate exceeded the organisation's standard of 75% for most of the day, with peaks up to 99%. These results confirmed the users' perceptions of crowding. The occupancy varied between floors.

The occupancy of enclosed workspaces was higher than the occupancy of workstations in open spaces. On average, 23% of the seats were temporarily vacant, with users being away but leaving their belongings. The proportion of deduced occupancy rose with increasing rates of actual occupancy and peaked around lunchtime.

The survey data provided insights into the average user perception of the new working environment. Despite the substantial length of the survey (93 questions), the response was high (59%). Compared with the average scores of flexible offices in the CfPB database, the users of this office were clearly less satisfied. The general issues regarding flexible offices, noise and lack of privacy were even more prominent in this office. The same applied to satisfaction with indoor climate, which often aligns with general dissatisfaction. In the group interviews, users complained about cold and drafts, for example, around the atrium, and stuffy meeting rooms. Another source of complaints was the uncomfortable furniture in the lounge areas, which was considered unsuitable for working in several respects.

Notably, the opportunities for communication and interior design, which are usually among the best-rated aspects of flexible (and recently renovated) offices, were rated as "poor" by the majority of the users. Despite the glass façades that allow for plenty of daylight, the average satisfaction with light in the workspace was relatively low, which may be due to the omnipresence of light-absorbing black

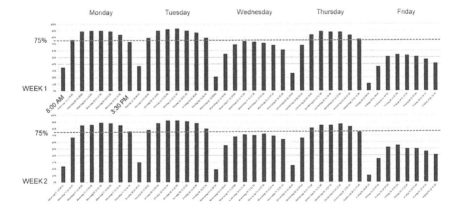

Figure 10.5 Observed occupancy rates (average of all working floors) for two weeks

in the interior. The widespread use of black and the general lack of colour also reduced satisfaction with the interior design. In the Netherlands, black has connotations of being either stylish or depressive: in the group interviews and remarks in the survey's text entry option, the latter association dominated. The users also explained that they experience the environment as monotonous, chilly and impersonal, although the modern style in itself is appreciated.

The standardised survey included four questions about satisfaction with non-physical aspects of the working environment (organisation, job, participation and remote working) to enable correction for dissatisfaction with the physical environment. In this case, satisfaction with job content is higher than elsewhere, and the group interviews confirmed the impression that most employees were highly engaged and motivated. The interviewees explained how the shortcomings of the physical environment, and especially the crowding and resulting problems, significantly reduce their job satisfaction, increase fatigue and make them reluctant to hold meetings with business partners. They feel that the crowded massive building, and desk-sharing policy, negatively influence social cohesion and knowledge-sharing within teams.

The building inspection revealed that, in general, sustainability and architectural standards were met but lacked user-friendliness. Positive scores included energy efficiency, spatial flexibility, safety measures, soundproofing, daylighting, quality and accessibility of coffee corners. Shortcomings included a lack of clear wayfinding, visual privacy, sound absorption within the building, plants, ambient lighting, opportunities for team personalisation, adjustable furniture arrangements in meeting rooms, low light levels, and lack of expression of organisational identity.

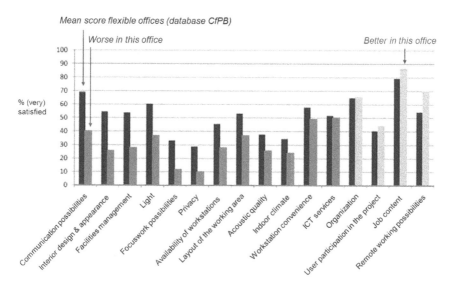

Figure 10.6 Benchmark of user satisfaction

Conclusion

The main conclusion of the POE was that the office building underperformed regarding usability and user experience, despite the available standards and considerable effort to involve users in the project. Evaluation of the process was not part of our assignment, so we could only speculate about what had gone wrong. The triangulation of research methods built a clear case for improvement. The report provided 42 interrelated recommendations for adjustments to the interior design, better alignment of services to the intensified use of the building, and guidance on user behaviour and management. Several improvements had been applied before Covid-19 struck. Since then, increased working from home has solved issues of severe crowding.

Case study C: Getting better all the time, Slovenia

Mark Eltringham   Alja Ceglar and Bostjan Erzen
Workplace Insight, UK   Kragelj Architects, Slovenia

Introduction

While the nature of work has already changed in many ways, the pace of change has increased even more dramatically over recent years. The challenge for designers and occupiers is how best to manage change, keep costs down and provide an adaptable home for the organisation. The standard answer to the challenge is to build flexibility into the building. If we take an idealised view of the modern office as a flexible, social space for a peripatetic, democratised and technologically literate workforce, the solution lies in increasing the use of desk sharing, drop-in zones, breakout space and other forms of multi-functional workspaces. In many offices, individual workspace is already being rapidly replaced by other types of space, quiet rooms and collaborative areas.

Flexibility must be hardwired into the building at a macro level. Not only must floorplates be capable of accepting a wide range of work styles and planning models, but servicing must also be appropriate and anticipate change, not just in terms of technology and telecoms but also basic human needs, such as having enough toilets to deal with changing occupational densities. It also means specifying the heating, ventilation and air-conditioning systems that can deal with the changing needs associated with different numbers of people and different types of equipment.

Elements of the interior that were once considered static now also need to offer far larger degrees of flexibility, particularly in terms of furniture, lighting, storage and partitions. This issue of flexibility has become more important within interior

Figure 10.7 Better's offices (photo: Matej Kolaković)

design. Interior elements should now define space, portray corporate identity, comply with legislation and act as an aid in wayfinding. They must do all this and be able to adapt as the organisation changes.

The past few years of unprecedented change in the way we work have made adaptability and agility key features of office design. A perfect example of this is provided by the headquarters of Better d.o.o., a Ljubljana-based technology company that provides healthcare solutions to more than 150 clients across 17 countries and securely supports over 30 million patients on its Better Platform.

Better by design

Better is part of a global community of healthcare practitioners, service and technology providers with shared values and a commitment to share information, technology and thinking on the best ways to deliver healthcare solutions. Better offers clients a digital health platform that is based on open data. It helps healthcare professionals to develop better solutions to simplify the work of health and care teams, as well as to improve the lives of patients.

"I believe healthcare should think about data first, applications second", explains Tomaž Gornik, founder and CEO of Better. "Separating data from applications prevents vendor lock-in and increases innovation. It enables healthcare organizations to move to an architecture built around a vendor-neutral, consistent, longitudinal patient record that serves as a single source of truth".

It is an organisation focused on applying a similar approach to its own business model. Decisions are driven by data, but the organisation also focuses on the wellbeing of its employees, democratic decision making and the need to constantly evolve as new information and technology become available.

The rapid growth and development of new markets around the world meant that Better had to move to a new purpose-built headquarters in Ljubljana. A projected growth rate of over 10% a year and over 130 staff made the move a crucial part of the company's strategy. The rapid growth and new expectations of flexible working cultures also made it essential that the building was able to accommodate change and growth effortlessly. It was also important that the design of the new offices reflected the company's values.

The building was based on an activity-based working model, which offers employees a choice of where to work based on their personal preferences, the tasks they are working on, the people they wish to work with and the need to come together organically as teams to work on new projects. The design also encourages people to move around the building, which is good for both their mental and physical wellbeing, including granting them more access to daylight and the naturally ventilated spaces of the building.

This kind of design is also better able to accommodate growth and change than a building based around dedicated workstations and rigid, hierarchical space and specification standards. It balances the needs of the business at a macro level while providing a tailored and constantly changing experience for each individual. This is also important in recruiting and retaining the often young professionals that drive the company's success.

POE I 2020

POE II 2022

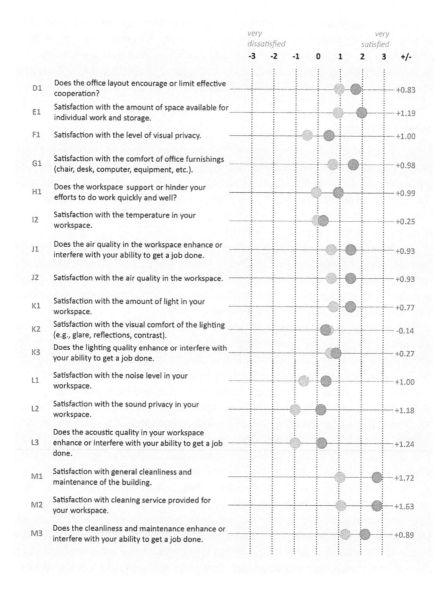

Figure 10.8 Occupant survey results 2020 versus 2022

The company chosen by Better to make this a reality was Ljubljana-based design and architectural practice Kragelj. Not only did the firm have an impressive track record of creating similarly sophisticated offices, but it also used an analytical approach to developing solutions that chimed with the client. The journey began with a series of pre-project evaluations and other analyses of the way the company worked, how information flowed, job roles, tasks and so on. Kragelj also carried out a number of utilisation studies to help determine space requirements in terms of both the size and the nature of specific types of space. These studies also helped to determine the needs of individuals and how these might change from day to day and over the longer term.

What was important in all of these studies was to create an office that could meet the business's current needs but also anticipate its growth and development over a longer period. This hardwired agility was to usher in a new type of collaborative working culture. This included finding ways to facilitate interactions and communication between all seven storeys and two basement levels of the 3,000 m² headquarters. Openness was essential not only on each floor but between them too.

This is where the pre-occupancy analytics and utilisation studies came into their own: the data enabled the designers to create a building with a dynamic that reflected the organic structure of the organisation. While each floor is a good example of the activity-based working model with a range of spaces and facilities, including a breakout space, private and quiet spaces, as well as team areas and individual workstations, each floor also has a sense of connection to the others, to encourage people to move around the building. The reception area is on the second floor and its design gives visitors a first glimpse of the thinking behind the entire building. As well as meeting spaces it incorporates a library with deck chairs to encourage people to read or relax while looking out at the world.

The building is also defined by one thing it lacks – a dedicated executive suite. Instead, everybody works on the same basis. The ground floor features a large shared space incorporating a restaurant and a choice of meeting spaces, presentation spaces, learning zones, relaxation and massage rooms. This ensures an informal flow of information within the organisation, facilitates moments of serendipity and helps to break down any silos that the structure of the building may have imposed on the organisation. Wayfinding is aided by the intelligent use of a dedicated colour palette on each floor.

The first floor has a tiered amphitheatre for get-togethers and presentations as well as performing a role as breakout space.

Getting better all the time

The sense of continuous improvement of the space is perhaps most evident in the sophisticated approach to its POE process and its use to drive an evolution of the design of the building. At the heart of this process are a series of ongoing questionnaires to determine how the building is delivering on a range of factors over time. This is essential, not only in determining how well the building is performing according to its occupants, but also in identifying where it can be improved.

The full results for the immediate post-occupancy and results for 2022 are shown in the figures, but in summary, there was an overall rise in the already excellent reported satisfaction levels with the building (98.7%). This was up by just over 14% from the initial POE.

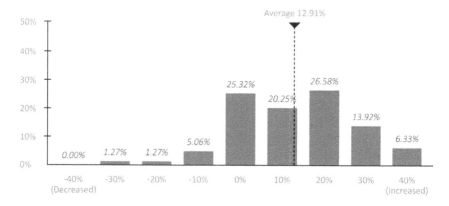

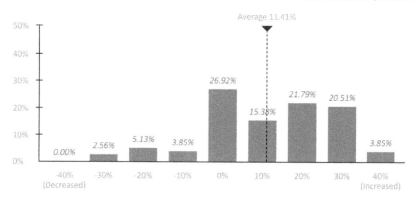

Figure 10.9 Post-project self-assessed productivity

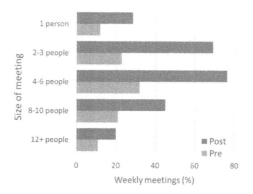

Figure 10.10 Pre- and post-project weekly meetings

Most notable in this regard were factors such as noise levels, office layout, communications, availability of appropriate spaces, temperature, air quality and the cleanliness and maintenance of the building. Many enjoyed near-perfect scores.

The biggest turnaround was seen in perceptions of acoustic privacy. In 2020, nearly two-thirds (63%) of people were dissatisfied with the acoustics of the office. By 2022 this had flipped completely with just under 62% of people now satisfied with their level of acoustic privacy.

The survey did indicate some room for improvement in certain factors, although only one – lighting comfort – had a lower score than in 2020. While it could be argued that providing the right temperature for everybody is extremely difficult to achieve, opportunities for areas to improve include both visual and acoustic privacy and certain aspects of the lighting design.

The survey also included responses to the question of whether people thought that both the office facilities and its environment made them more or less productive. A little under 8% of people felt the facilities made them less productive, a quarter felt they made no difference, with the remainder saying they had increased their productivity. Overall, the average increase in perceived productivity was just under 13%. There was a similar distribution of responses to the question of how the office environment affected productivity. Slightly more people felt dissatisfied with the impact on their productivity, but the overall average increase in reported productivity was still over 11%. This may be a reflection of the points about lighting revealed by the survey.

Changes in the levels of collaboration between 2020 and 2022 were measured both qualitatively and quantitatively. This is seen as one of the key success indicators for the office. People expressed near-universal levels of satisfaction with the availability and provision of meeting spaces in the building. This, in turn, suggests that the analytics and studies used in developing the design of the building had been more or less spot on. This is reflected in terms of the generally increased number of meetings that now take place. Between 2020 and 2022 the number of weekly meetings taking place jumped significantly across a range of meeting types. The proportion of people saying they took part in weekly meetings more than doubled, and in some cases nearly trebled, regardless of the numbers of people involved. Changes in the number of daily and monthly meetings were more consistent, although there was a fairly significant reduction in the number of daily meetings involving 2–3 people, which may have an impact on future designs.

Conclusion

The results from the Better headquarters building show the value of pre-occupancy surveys and analytics in developing an appropriate design for an organisation and its building and people and thus delivering post-occupancy success.

There will always be room for improvement; indeed, the ability to adapt in this way and in response to the growth and evolution of a business is essential. The improvements in levels of satisfaction, productivity and collaboration between 2020 and 2022 highlight the crucial role a continuous process of POE can play in improving the performance of an organisation as well as delivering important well-being and productivity benefits for individuals.

Case study D: Evaluating a workplace intervention study, Finland

Piia Markkanen

Oulu School of Architecture, Faculty of Technology, University of Oulu, Finland

The impact of the pandemic on offices

The Covid-19 pandemic disrupted the ways of working in such a manner that, upon returning to the offices, it became clear that many spaces and guidelines for designing working environments no longer fit employees' needs. The pre-pandemic studies predicted decreasing desk space demand and increasing space demand for collaboration, meeting spaces and social areas.[3] Currently, new challenges are forcing designers to rethink office design to support hybrid work and create more inviting collaboration and social spaces to maintain organisational cultures. At the same time, there is a need to reduce space costs, partly due to the widespread adoption of remote work.

The work environment provides surroundings that have the potential to support employees' wellbeing, satisfaction, work engagement and motivation. However, as design processes define the future use of the office space, they need to be informed with timely and organisation-specific information.

Participatory design approach

Work environment research studies generally focus on existing work environments or organisational change processes initiated by companies and organisations. Our research project provided an interesting yet challenging opportunity to understand

Multifunctional workspace   Breakout area

Figure 10.11 Multifunctional workspace and breakout area after design intervention

the impact of workplace design in a local company that participated in our research study. In this case study, we approached the work environment design challenges with research-by-design methods using a participatory design approach[4] to design a workplace intervention study. The approach utilised the need – supply fit model and a multidimensional design framework during the design and evaluation phases. Need – supply fit is a sub-theory of person – environment fit theory, and it describes how employees' needs are supplied by the work environment.[5]

The fit formation is often assessed by how the surroundings support activities with low or high task complexities and the needed level of privacy. In our research, we extended the need – supply fit model to include the need for interaction and the need for atmosphere. The multidimensional design framework consists of instrumental, aesthetic and symbolic dimensions. Together they form the comprehensive spatial design that supports the space's functional use and gives the space its characteristic features. The aesthetic and symbolic dimensions influence the spatial atmosphere, the sensory experience and, more importantly, convey the intended use of the space. The concept of "atmosphere" is part of our methodological setup because it provides an understandable way to communicate the design features of different spaces with study participants.

The subsequent design process aimed to increase employees' workplace satisfaction. Work environment researchers use the definition of workplace satisfaction to describe how well the physical work environment supports its users' needs. Different office types impact employees' satisfaction through the layouts and structures that protect or allow either privacy or exposure to distractions.[6] It is important to note that the experience of workplace satisfaction is an outcome of perceived comprehensive design, indoor qualities, and also personal experiences. The research knowledge of workplace interior design is inadequate because of the

Formal meeting room   Informal meeting room

Figure 10.12 Meeting rooms area after design intervention

lack of suitable methodology for analysing and communicating the detailed level of design.

Pre-design process and post-intervention evaluation

POEs are typically performed after 6–12 months of occupancy. While our research project set limitations to the study's timeframe, it opened possibilities to explore different quantitative and qualitative research methods. Combining the pre-design, design and post-design phases in the same study gave us an opportunity to link the design data to the evaluation process and to communicate the research objectives with the study participants.

Altogether, 15 employee volunteers participated in the pre-design process. The intervention's focus was on enhancing workplace satisfaction through improved symbolic, aesthetic and atmospheric characteristics. The context of the research in this case study was the headquarters of a health technology company with 50 employees working in the local office. The office floorplan displayed the characteristic features of an activity-based office sub-type, the combi-office. Assigned workstations were located in private offices and multitenant workspaces. In addition to the various meeting rooms and breakout areas, the office provided small-scale production and testing spaces. The research area contained a breakout area and three shared workspaces: a multifunctional workspace for quick meetings and individual work; a formal meeting room for board meetings and onsite visitor meetings; and an informal meeting room for team meetings, product development and brainstorming. The spaces had established uses and functional setups, which were not changed.

The pre-design process included a participatory design approach: the employees were first interviewed to gain detailed information on their job roles, daily activities and activity-related needs, such as privacy, interaction and online work. Next, the employees were invited to online participatory design workshops, during which they explored activity-related needs, preferences, feelings and activity-supporting atmospheres. The workshop results were analysed and fed into designing the workplace design intervention. The intervention changes included new lighting, curtains, furniture, drawing boards and acoustic elements.

Experience sampling method

Our multidisciplinary research team tested the following evaluation methods in this case study: a wellbeing survey (online, quantitative), an experience sampling method (on-site mobile, quantitative), physiological measurements (on-site, wearable device, quantitative), semi-structured interviews (on-site, qualitative), and evaluation workshops (online, qualitative). The studies were conducted with nine employee volunteers within three months of completing the interventions. The small number of study participants and remote working due to Covid-19 limited the data collection. The most suitable methods with the limited number of participants were the experience sampling method (ESM) and on-site interviews.

The rest of this case study explores the potential of the ESM as a POE method in more detail. ESM can be used to study implemented changes in workplaces, explore how spaces are used and experienced and investigate immediate experiences in a natural environment. The collected data can inform how employees' workplace satisfaction and need – supply fit are formed on spatial- and organisation-specific levels. We set up a dynamic ESM that collects repetitive data samples of work- and recovery-related situations.[7]

Fundamental to ESM is the collection of repetitive and representative sampling to gather information on multiple experiences of the same situations. These days the reports are often collected using mobile devices, such as smartphones or wearable electronics, so when designing an experience sampling query, it is essential to focus on how the sample is to be collected to gain targeted information about the experienced phenomena. Different sampling approaches include interval-contingent (experiences are reported at regular timeframes), event-contingent (where the study participant reports when a particular event occurs), and signal-contingent, with either fixed or random signals as a reminder for the user to report the experience. The signal-contingent ESM enables gathering contextual information in office spaces.

For this case study, we built a research setup consisting of a location positioning system (Noccela) to determine participants' real-time location within the research area and the ESM system. The ESM system was designed to detect participants' movement between the research spaces and to send a notification when the following criterion had been fulfilled: participants exited a research space in which they had spent a minimum of 20 minutes and less than 10 hours. Setting this criterion enabled us to focus the queries on meaningful events. When the criterion was met, a notification request was sent to participants' smartphones through an application controlling interface and iOS mobile application developed for this case study. The mobile application pushed the signal to study participants and served as a questionnaire platform. We defined this method as "dynamic experience sampling" because the signal depends on how participants use their work environment. Notably, the setup collects space-specific information.

Preliminary findings

The ESM query should be prompt and straightforward to fill. We requested nominal information on the participant's location (assigned workstation, other workstation, multi-function workspace, formal meeting room, informal meeting room, breakout area, and other location) and activity (working alone, working together, working together on the phone, working together on videoconference, recovering alone, recovering together, other activity). The task complexity, need for privacy, need for interaction, need for atmosphere, and experienced spatial support was questioned using a Likert scale (1 to 5). The task complexities were higher for videoconferences and collaborative activities than for individual work. Interestingly, collaborative activities increased the need for privacy, interaction and atmosphere. The level of privacy required was higher in meeting rooms than at assigned workstations. However, the videoconference meetings increased the privacy needs at the workstations. The recovery events required more versatile spaces than this case study company had to offer. The group recovery situations were well supported in the breakout room, but the individual recovery events had no allocated spaces. Thus, these events were reported in the "other" location category.

Overall, all intervention spaces gained higher spatial support scores during the intervention. Significantly, this ESM study assessed the importance of spatial atmosphere. The spatial quality impacts the overall experience and satisfaction towards the space. A supportive atmosphere was needed in collaborative work activities and group recovery events. Overall, these findings imply the need to pay more attention to the design of meeting rooms, recovery areas and spaces dedicated to collaboration.

Current changes in working practices set new and emerging needs for knowledge work environments: agile and contextual evaluation methods are needed to respond to the changing needs in existing workplaces. The ESM can be set up in more simplified ways than in our research study. The collected information requires qualitative data to inform research and design processes, but depending on the inquiry, the ESM may provide an accessible way to update understanding of changing requirements in work environments. Overall, the design processes should be informed by organisation-specific and contextual information in order to develop work environments to support different individual, group and organisational activities in addition to recovery and restoration during the working day.

Acknowledgements

This case study presents partial results of the "ActiveWorkSpace – Spatial solutions for environmental satisfaction, wellbeing and work engagement in activity-based offices" research project funded by the Academy of Finland (grant number 314597). The intervention research study was a collaboration of research teams led by Aulikki Herneoja (University of Oulu) and Virpi Ruohomäki (Finnish Institute of Occupational Health).

Notes

3 Harris R. (2016) New organisations and new workplaces: Implications for workplace design and management. *Journal of Corporate Real Estate*, 18(1), 4–16.
4 Bannon Liam J. and Ehn Pelle. (2012) Design matters in participatory design. In Jesper Simonsen and Toni Robertson (eds) *Routledge Handbook of Participatory Design*, 37–63, London: Routledge.
5 Kristof-Brown A.L., Zimmerman R.D. and Johnson E.C. (2005) Consequences of individuals' fit at work: A meta-analysis of person-job, person-organization, person-group, and person-supervisor fit. *Personnel Psychology*, 58, 281–342.
6 Heerwagen J., Kampschroer K., Powell K. M. and Loftness V. (2004) Collaborative knowledge work environments. *Building Research & Information*, 32(6), 510–528.
7 Markkanen P., Paananen V., Hosio S. and Herneoja, A. (2022) Dynamic experience sampling method for evaluating workplace experiences. In C. Tagliaro, A. Migliore and R. Silvestri (eds), *Proceedings of the 3rd Transdisciplinary Workplace Research Conference*. Milan: Politecnico di Milano, 599–610.

Case study E: Evaluating adaptive re-use: Former UK carpark transformed into work and learning space

Ziona Strelitz
ZZA Responsive User Environments

Background

POE is 'the core expertise of ZZA, a group of design anthropologists who use occupant feedback at the heart of the design process. It has an extensive repertoire of evaluations involving new build and adaptations, and covering varied typologies and scales, interior fit-outs, buildings and campuses. Most of this research has been commissioned by developers and occupiers – particularly clients looking to inform subsequent procurement with evidence and insights from projects they have already delivered.

There is still an expectation for architects to initiate an independent evaluation of their outputs, and Make is distinctive among architectural practices for commissioning specialist evaluations across a range of its projects to learn how the designs work for users. In 2018, ZZA evaluated Make's commercial groundscraper at 5 Broadgate, London, followed in 2019 by a POE of the practice's award-winning Teaching & Learning Building at the University of Nottingham. Both of these studies were entirely unfettered by the architects. In 2022, Make exposed the practice to even greater scrutiny by commissioning ZZA to undertake a POE of its own studio, which it had designed and occupied since 2015. That evaluation is the subject of this case study.

The site

An extreme example of adaptive re-use, Make's studio occupies a large open plan lower ground level space – the former carpark of an office building. By creating this work environment, Make could accommodate all of its 130 people on a single floor in a central London location. The transformed space retains the legible features of the former car park. The studio is entered via the ramp that was previously used for vehicle access; the columns still display the co-ordinates that denoted where cars had been parked. Further specific conditions of the legacy space include the deep plan, with its advantage for organisational unity and its limitation on natural ventilation, and a relative lack of external aspect and natural light. There are, however, varying levels of daylight at all four edges of the studio floor, as well as light that comes through a section of glass ceiling blocks. And there is an external courtyard with planting and informal furniture, which is visually and practically accessible from the studio interior through large glass windows and doors.

With the current drive for net zero carbon in the built environment, and the impetus to use rather than demolish existing built stock, the evaluation of this studio has special relevance. How effective is such an extreme transformation in use? If there is clear evidence that it is working for its occupants, that gives promise for other imaginative and purposeful adaptations of use. And to the extent it generates pointers to enhance future performance, that is the learning benefit of commissioning such evaluation.

The POE

Seven years is a benchmark for the life of a fit-out, and evaluating Make's space at this juncture affords a perspective on the interior as well as the base space, reflecting any change in expectations and working practices that have ensued since the original move. In this case, the timing also captured evolution in work modes arising from the Covid-19 pandemic.

The method followed ZZA's established practice. The core comprised a set of questions systematically addressed, one-to-one, in scheduled face-to-face interviews with a sample of 20% of the studio's occupants. The interviewees represented a wide range of roles across design and business support and a spread of experience spanning director to assistant roles. The studio's accommodation is arranged as a field of spatial clusters, and the sample included a contributor from all of these, thereby enlisting user perspectives that represented the range of internal conditions at play in the studio's plan and spatial layout.

Given the timing of the POE – when the studio was "regrouping" in person after Covid – it was judged appropriate to open the study beyond the sample to any occupant who wanted to share their experience of the space in a walk-up clinic. These sessions, too, were conducted on a one-to-one basis and were documented separately.

Scope

The POE content covered the studio as a physical space, its role in the life of "Makers", its effects on work, professional development, and meeting strategic organisational goals.

The research questions referenced the studio's component aspects and spaces:

- identity/image,
- access and arrival,
- circulation and navigation,
- spatial quality,
- space for solo and group work,
- air,
- light,
- acoustics,
- amenities,
- WCs and showers,
- technology,
- flexibility,
- client hosting,
- look and feel,
- overall assessment.

In total, this involved 102 evaluative questions supplemented by questions on people's patterns of use (e.g. travel mode, typical duration in the studio, and lunch venue). Evidence of such usage is important in understanding the nature and context of occupants' experience of the space. The overall coverage affords a full picture of the space in use for people who work there. The face-to-face method enabled thoughtful, considered responses, discussion and clarification, and hence a high degree of validity in the outputs generated by the POE relative to the meaning that interviewees intended.

Outputs

ZZA's outputs are quantitative and qualitative. The metrics are generated by the evaluation codes that interviewees select for each question; the qualitative content derives from their narrative comments. The latter is important in conveying the reasoning and sentiment that underly users' ratings; the two evidence streams are complementary.

The aggregated coded responses are classified by a high bar of user satisfaction as follows:

- major successes: 80%+ evaluate a codable aspect as "positive",
- success: 80%+ evaluate a codable aspect as "positive" or "ok",
- issue: >20% evaluate a codable aspect as "negative".

Snapshot of Make's studio, seven years on

The POE's major successes include important endorsements of the studio's adapted space, as well as its support for organisational performance and occupant satisfaction:

- demonstrates vision in adaptive re-use of a building: Positive 96%,
- showcases a building adapted for workplace use: Positive 88%,
- looks and feels professional: Positive 80%,
- facilitates communication across Make's teams: Positive 80%,
- facilitates a sense of community: Positive 89%,
- pleased with the studio environment overall: Positive 81%.

Across the full set of POE questions, a high majority of user evaluations were positive.

The occupants' endorsement of the building is further evidenced in the following contrasting profiles. Asked to cite their "favourite thing about Make's physical studio":

- 100% cited at least one favourite thing,
- 27% exceeded the question by citing more than one thing,
- no one cited nothing.

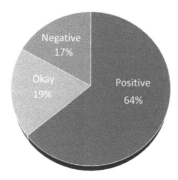

Figure 10.13 Profile of interviewee evaluations

Asked if they had an "Unfavourite thing about the physical studio":

- 35% had no "unfavourite thing",
- 58% cited one thing,
- 8% cited two things.

Convergence of work and learning

Although the studio was primarily conceived and designed as a workplace, most interviewees were also positive about the studio as a learning environment:

"I learn massively here . . . "

"It feels more of a working studio, more of a student style layout – very informal."

"The disciplines are quite close to you – you can see what they're doing."

Given the changing nature of work and the qualities that contemporary cohorts aspire to achieve in their work activity, these benefits that users perceived as deriving from the design are important.

Interviewees pointed to the range of interior settings to support their view that the studio "Showcases contemporary ways to work and learn":

"The mix of places. The fact that we've got different places to work – like the breakout and kitchen, the central area for sketching, the meeting rooms which *aren't* usually closed so you can see in, and the ramp."

"It's how the circles [of desks] work, and the higher desks – you can talk to people at the same height: you don't have to tower over them if you're standing talking."

Cultural memory

Make's people value the overt way that the studio signals its former use as a car park. The legacy is announced on arrival at the studio, and as one proceeds down the entry ramp, with its exhibition of displays on one side, tables for team collaboration on the other, and beyond that, a dramatic view of the floor plate and the teams in action. The POE evidenced people's appreciation of this expressive value as distinctive and special, with the studio's legible legacy registering prominently in users' favourite things about the studio:

"The whole story of the adaptive re-use – I always tell it to clients and students."

"The ramp – the sense of procession and openness."

"The fact that it's an old car park, and not sectioned off – not 'officed off'."

"It's the only place I've ever worked, but I think it's quite unique. I haven't seen another in a car park."

"It just has a unique character based on its unusual location in a car park. Most [organisations] of our scale would have filled out an office building."

Users welcome the design's incorporation and celebration of the studio's "found" space. Their narrative comments reflect on the characteristics of the space and also on the benefits of the evaluation methodology that is oriented to identifying why "good" aspects are perceived as beneficial. This is learning about building performance that cannot be acquired by users ticking boxes:

"The converted building retained and adapted features which give it character – it's the opposite of a fit-out."

"It's the character of the space. It's been designed around the existing features – the columns, exposed brickwork and concrete."

"It feels vibrant – like a workshop; collaborative, making and doing – not sterile."

"The openness and the honesty to the building structure and services. And it's playful."

Visual porosity

Although the design's openness is physical, the interviewees perceive it as the basis of the social inclusion and organisational engagement that they feel:

"The *openness* – no barriers."

"It's *open* – you see everyone every day."

"Everything's usable and I don't feel shy to use any part of it."

"The open plan. It's really nice being able to see everyone on the same level. It's great for communication."

"The buzz, the sense of excitement. It's a result of its openness and integration of everyone in the same space."

"The sense of openness – physical and social."

"When you're in the space you feel in the heart of something bigger – being in the heart communicates what Make's about – the 'Cathedral of Make', surrounded by our values and what we're about and it's visible spatially and culturally."

Situational challenges

Alongside its engaging iconography, a lower ground former car park poses challenges relating to thermal experience, daylight and external vistas, which the POE identified. Crucially, as a case study to help inform other adaptive schemes, all the issues associated with these conditions are addressable, and relevant interventions could be effected where owner-occupancy or suitable lease terms are facilitated.

Loose-fit context

The POE generated user endorsement across a range of infrastructural provisions that support day-to-day studio operation. There were major successes in terms of workstation furniture, breakout spaces to relax in, WCs, access to showers, plants and various aspects of technology. The learning point is the influential role of small moves in shaping positive outcomes.

Social change

The evaluation also signalled a fault line, catalysed by the ubiquity of virtual meetings heralded by Covid-19 and the wide incidence of hybrid collaborations and events since the pandemic: namely, where online participation is conducted in an open plan work environment, individuals' own loud speech tends to be muffled by their headphones, but it often disrupts colleagues working around them.

This important POE finding exemplifies the productive learning that POEs can identify, flagging up disconnects arising from social and technological change that require productive solutions.

Stewardship

ZZA's Make studio evaluation reinforces findings from other POEs by showing that the nature and quality of facilities management directly impacts users' feelings about their environment. Here, all aspects of facilities management were evaluated as major successes:

"Effective balance of welcome and security": Positive 92%,

"All parts of the studio appear and feel clean": Positive 85%,

"Studio appears generally well maintained": Positive 89%,

"Provision for waste collection seems effective": Positive 89%.

Answering this POE's central question

All users mentioned daylight, external views and thermal experience 'when asked about aspirations for the physical studio, but not more so than loose-fit aspects connected with layout and amenity, support infrastructure, and more agility in the studio's use. This further evidence that the studio sitting located on the lower ground floor is not an overriding determinant bodes well for the prospects of other ambitious transformations.

Case study F: Evaluation of an English primary school

Gary J Raw
GR People Solutions, UK

Introduction

This POE was conducted in a recently constructed primary school (for children aged 4–11) in the south-east of England. The request came from the school as a result of staff complaints about overheating. The method was multifaceted, but I focus here on what was achieved with a questionnaire survey. Two key aspects were to pilot a novel set of questionnaires and to include all occupant types – teachers, other staff and students. I report some selected aspects of the work to demonstrate the approach; there is additional content in Raw (2022).[8]

Method

The survey was conducted on a warm day in early September. In addition to staff, it included the four oldest year groups of students (ages 7–11), classified in England as Key Stage 2 (KS2). There were eight KS2 classes, two in each year group, each of approximately 30 students. In each year group, one class was on the south side and the other on the north side of the same floor of the building; this made it possible for comparisons between warmer and cooler spaces.

The classrooms have large glazed areas along one façade, with manually operated internal shading by horizontal blinds. There are openable windows plus supply-only mechanical ventilation (MV). The MV delivers unconditioned outdoor air from a single outlet near the front of the room at ceiling level. Supply air is directed towards the rear of the room, initially flowing close to the ceiling. Large low-speed overhead fans were installed in three of the KS2 classrooms to evaluate them as a remedy for overheating.

As in many buildings, the thermal conditions inside the school were expected to vary widely – between days or seasons and within a day. The occupants' physical activity and clothing also vary markedly, sometimes within hours or minutes. A limited range of variation should not be problematic (it can reasonably be argued to be beneficial). But there is the limit to the variation – especially rapid variation – that can be managed or tolerated (even if good adaptation can be achieved to each individual condition).

Because of this variation, the common approach of using a seven-point scale of thermal sensation would have had limited value. In order to get meaningful results, it would have been necessary to repeat the survey multiple times during many days, across seasons. I was also aware that design standards are often expressed in terms of the percentage of the time that conditions fall outside a recommended range, not the conditions at a point in time. Therefore, I developed new questionnaires more appropriate to the varying conditions.

The questionnaires asked about thermal conditions and control over them, indoor air quality (IAQ), light and noise. There were two questionnaires for staff, one to report on conditions during the previous summer and winter, and one to report on conditions on the day of the survey. This report covers only the seasonal questionnaire. Student questionnaires covered only conditions on the day of the survey. Draft questionnaires were first subjected to cognitive testing with volunteers from other schools (primary school teachers and students within the target age range). Space does not allow a complete account of the designs, so I focus the description on the parts that are most relevant to the reported findings. Some more detail is given later, along with the findings.

The staff questionnaire included, separately for summer and winter, questions about:

- the percentage of time when it is (a) too cold and (b) too warm in the respondent's usual space,
- aspects of IAQ, using seven-point scale ratings.

The student questionnaire used an age-adapted seven-point thermal comfort scale: "How do you feel about the temperature in the room at the moment?" Response options were: "Much too cool", "A bit too cool", "Cool but comfortable", "Just right", "Warm but comfortable", "A bit too warm" and "Much too warm". For analysis purposes, these responses were coded -3 to +3. Students were then asked whether they remembered any of the 11 things happening during that day (yes or no), including: "Feeling too hot", "Feeling too cold", "Cold air blowing on you", "Feeling that the air in the class is dry or dusty", "Unpleasant smells", "The electric lights being too bright" and "Sunlight in your eyes".

Student questionnaires were coded to identify their location within each room, defined by whether the student was (a) nearer versus further away from the windows and (b) nearer the front of the room versus nearer the rear.

Results and discussion

Table 10.1 shows the percentage of time that staff reported being too cold or too warm. Overheating was, as expected, more frequent in summer and on the south side but also experienced the majority of the time on the north side in summer. However, classrooms could also be too cold, particularly on the north side, even in summer. In each season, it could be sometimes too warm and sometimes too cold in the same room.

Consistent with these findings, seven-point ratings of "Still – Draughty" were higher (more air movement) in winter than in summer (3.4 vs 2.0) and higher on the north side than the south (4.0 vs 3.3 in winter, 3.1 vs 1.3 in summer).

Table 10.1 Mean percentage of time that staff report being too warm or too cold

Side of rooms	Summer		Winter	
	Too warm	*Too cold*	*Too warm*	*Too cold*
South side	77	0	25	24
North side	61	21	15	77

Staff tried to control the temperature (mainly using thermostats, windows and blinds) but reported that they did not have effective control. Some had brought in heaters or fans; some wore a coat in winter while teaching.

The temperature can affect both the actual pollutant concentrations and the perceived IAQ at a given concentration. It is, therefore, consistent with the thermal comfort findings that IAQ was perceived to be better in winter than in summer (even though windows are more likely to be open in summer) and better on the north side. Mean ratings of the staff are shown in Figure 10.14.

The students' mean thermal comfort vote was 0.6, "Warm but comfortable". However, only 58% of students were in the comfort range (middle three points). Furthermore, many students reported being too hot or too cold at some point during the day (see Table 10.2). As with staff, there was a clear difference between the north and south sides.

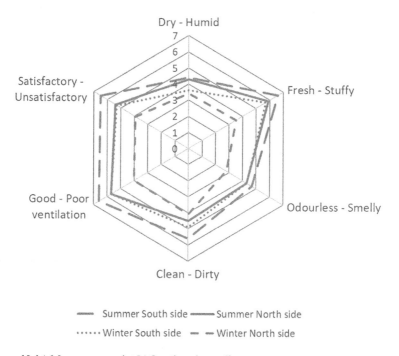

Figure 10.14 Mean seven-point IAQ ratings by staff

Table 10.2 Percentage of students reporting feeling too hot, too cold, or cold air blowing on them

Side of rooms	Too hot	Too cold	Cold air movement
South side	56	4	23
North side	31	31	37

More students on the south side than on the north reported "Unpleasant smells" (26% vs 14%), but fewer reported "Dry or dusty air" (25% vs 30%).

The student survey added to what could be learned from the staff survey by offering greater spatial analysis. This was especially important for understanding the effect of the ceiling fans. There were too few staff to draw quantitative conclusions about the fans although, informally, opinions were positive. Comparisons are made here between students in rooms with and without fans, on the south side only. Findings from the south side are more robust because (a) there were two rooms with fans and two without and (b) the rooms with fans were in the middle of the age range. On the north side, only one room had fans.

Students in rooms with fans were moderately less likely to have felt too hot (52% vs 60%). This difference was consistent between the front and rear of rooms, but other effects were not consistent (see Figure 10.15). Only students at the front of rooms reported more cold air movement in rooms with fans. This was reflected only weakly in reports of feeling too cold, suggesting that the air movement could be beneficial. Also, students at the rear of rooms had a higher (warmer) comfort vote with fans present (1.4 vs 0.9), but there was no difference at the front of rooms (1.2 with or without fans).

IAQ responses provide further evidence of the effect of the fans (Table 10.3). Improvement due to the fans was greater at the front of rooms and seen for both IAQ variables. Thus, IAQ was perceived more negatively at the front of rooms where there are no fans than at the rear of rooms where there are fans.

The improvement in perceptions nearer the front of rooms with fans (partly balanced by some worsening of perceptions at the rear of rooms) suggests that the fans are not simply increasing air speed and turbulence. The fans could be displacing

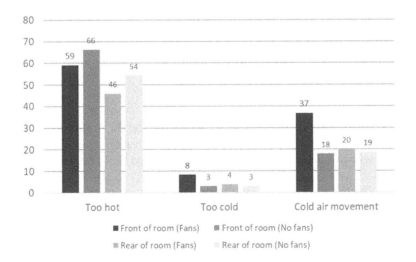

Figure 10.15 Percentage of students reporting feeling too hot, too cold and cold air blowing on them

Table 10.3 Percentage of students reporting dry/dusty air or unpleasant smells

Position in room	Dry or dusty air		Unpleasant smells	
	Fans	No fans	Fans	No fans
Front of room	9	49	18	46
Rear of room	16	27	23	19

Table 10.4 Mean percentage of students reporting problems with light

Position in rooms	Sunlight in your eyes		Electric lights too bright	
	Fans	No fans	Fans	No fans
Front of room	37	42	6	40
Rear of room	58	37	6	13

MV supply air downward, nearer to the front of the room. This could result in staff experiencing greater cooling than students: staff spend more time at the front of the room and they are taller and more often standing, therefore closer to the fans.

Students' reports of problems with light add further insight (see Table 10.4). Differences between the front and rear of rooms, with and without fans, could result from changes in teachers' behaviour: if they feel cooler with fans, they might raise the window blinds. This would increase students' exposure to sunlight, especially at the rear of rooms (where more windows are in the field of view). There would also be less need for electric lights to be on; this would benefit mainly students at the front of rooms, where the ceiling height is reduced, placing lights more in students' field of view. A reduction in window opening can also be hypothesised. Such changes in behaviour could partly account for why the fans had only a moderate effect on students' reports of being too hot.

Conclusion

This POE demonstrated a cost-effective triage method for evaluating an indoor environment. With further development, and the creation of a database from multiple schools (and other environments), the method could become even more powerful.

The subjective responses give a picture of the indoor environment and the impact of the ceiling fans that is coherent in itself and consistent with classroom characteristics. The survey confirmed the suspected overheating problem and described it in greater spatial detail, while also identifying additional problems and evaluating mitigation. The findings also suggest a logic to how teachers sought to manage classroom environments, with thermal comfort competing with other factors that influence how heating and cooling are managed.

Any further action should give attention to consequential changes in the use of windows and shading, and the balance of natural and electric lighting. The findings also indicate what follow-up should be carried out to confirm conclusions: objective measurements in various locations within rooms, not in a single location, and observations and discussion with staff to clarify their use of windows and blinds.

Acknowledgements

This work was funded by CETEC Foray Ltd, with support and advice from Paul Ajiboye (CETEC UK) and Richard Daniels (UK Department for Education). Andrew Bellamy and Jamie Liang of CETEC assisted with questionnaire distribution and collection. The kind cooperation of the staff and students of the school is gratefully acknowledged.

Note

8 Raw G.J. (2022) Varying comfort: A different kind of challenge. In S. Roaf and W. Finlayson (eds) *The 3rd International Conference on Comfort at the Extremes: COVID, Climate Change and Ventilation*. Edinburgh, Scotland, 348–357, 5–6 September. https://comfortattheextremes.com/september-papers/

Case study G: Evaluation of Lick-Wilmerding High School in the US

David Lehrer
Center for the Built Environment, University of California, Berkeley

Context

Investments in public elementary and secondary schools are often lacking, resulting in problems that negatively impact students and teachers. This situation is especially dire in lower-income communities, and this was made especially clear during the Covid-19 pandemic, which revealed common deficiencies in school ventilation and heating, ventilation and air-conditioning systems. In this context, an encouraging example can be seen in the renovation and expansion of a mid-20th-century high school in San Francisco, Lick-Wilmerding High School (LWHS).[9] This project by Esherick Homsey Dodge and Davis (EHDD) Architecture offers useful lessons on how to update obsolete schools to meet current criteria for learning environments and sustainability while preserving the building's historic fabric and improving connections to the local community. The project was also designed to provide net-zero energy performance with numerous sustainability features and has received several significant design awards. The POE of this renovation and expansion project was performed by EHDD using the CBE Occupant Survey, a resource managed by UC Berkeley's Center for the Built Environment.

Figure 10.16 This view shows the new corner entry with public-facing activities, the historic façade below far left, new classrooms on the second story and the surrounding urban context (photo: Michael David Rose)

School renovation programme and goals

The architectural programme was driven by the need to expand classroom and administrative spaces, create a sustainable and resilient campus, and improve indoor conditions, especially acoustics and indoor air quality (IAQ). The designers also wanted to create updated learning and public spaces that would promote interaction and social connection and support multi-functional uses for collaboration. They designed a variety of spaces to enable interactions among small to large groups within rooms and corridors, including openings to admit views and daylight. The new floorplans included areas for the display of student work, reflecting the school's legacy of teaching industrial arts. The new design also moves student meeting areas to public areas near the main entrance, including the school's Center for Civic Engagement, exhibiting the school's commitment to social equity and diversity.

Bringing an older school up to current standards has special challenges for designers because schools in California are typically built of light construction that does not provide the indoor conditions teachers and students now expect. At LWHS the local surroundings had become urbanised since the school opened in the 1950s, and consequently, the lightweight exterior walls and single-pane windows did not sufficiently block sound from the adjacent traffic, light rail and a nearby freeway. However, the architects needed to preserve the character of the 1950s façade, including its narrow window mullion profiles. This was accomplished through the design of removable acrylic panels installed inside

Figure 10.17 New classrooms on the upper level were designed for access to quality views and daylight (photo: Michael David Rose)

the original windows, thereby preserving the original character while providing acceptable thermal and acoustical performance. In addition, much of the extra space required under the redevelopment programme was provided by a third-storey addition, which was intentionally set back from the original façade. The new façade includes extensive glazing to provide access to quality views and daylight, but with varying densities of ceramic frit on the south and west exposures to control glare and heat.

Before the renovation, IAQ had been impacted from the adjacent freeway and occasionally by wildfire smoke events that are unfortunately experienced in California. The renovation addressed this through an updated mechanical ventilation system with a dedicated outside air system and a high air filtration minimum efficiency reporting value (MERV).

Building evaluation at LWHS using the CBE Occupant Survey

The results of this renovation and expansion were evaluated by EHDD in collaboration with the school staff using the CBE Occupant Survey, a web-based resource provided by UC Berkeley's Center for the Built Environment.[10] The survey tool was developed with funding from the US General Services Administration in the early 2000s and was initially used to evaluate the effectiveness of property management in hundreds of federal offices annually. Since that time the survey has been used in over 1,000 buildings as a post-occupancy tool. It is now offered as a benefit to CBE's industry consortium members and is provided to other design firms and commercial building owners for a nominal fee.

The standard questionnaire covers nine key aspects of the indoor environment: thermal comfort, air quality, acoustics, lighting, cleanliness, spatial layout, office furnishings, workplace in general, and the building in general. Most questions use a 7-point Likert scale, with additional open-ended (text box) questions, and many Likert scale questions include branching (conditional) questions in order to drill down into the possible reasons for dissatisfaction. The survey also captures information about each responder, including age, gender, workspace type and location, and the length of time working in the building and in their current space. A number of optional question sets are also available to gather feedback about additional topics. These were developed based on research needs or requests from survey users. Versions of the survey are available for several occupancy types and have been primarily used for offices, K-12 schools, halls of residence and healthcare buildings.

Questionnaires have also been tailored for numerous research efforts, for example, to study thermal comfort, acoustics, LEED-certified buildings, the effectiveness of K-12 schools and many other topics of relevance to commercial building stakeholders. By maintaining a consistent set of core questions, these survey activities have yielded a robust benchmarking database with responses from 90,000 occupants and over 900 buildings. In 2021 CBE researchers published a detailed analysis of this survey database and its implications for building designers and operators, and guidance for survey-based POE practices.[11]

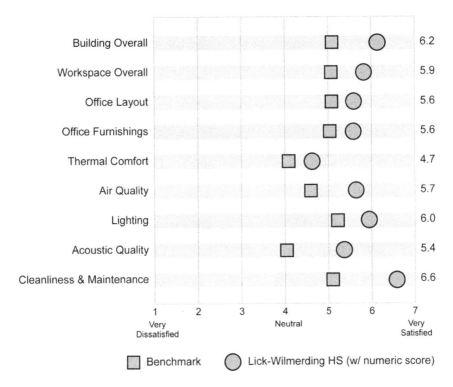

Figure 10.18 The benchmarking scorecard for Lick-Wilmerding HS shows that survey results for all categories compare favourably to the aggregate averages from approximately 900 buildings

CBE suggests that new or renovated projects allow a minimum of 6 months between move-in and a survey implementation, with 12 months preferred to allow occupants to experience a full year of seasons and annual activities. The LWHS renovation was completed by the end of 2018, occupied in January 2019, and the CBE survey was launched during October 2019. A link to the survey was emailed to the teachers and staff, and responses were recorded from 55 people from a possible pool of approximately 60, a very high response rate. Luckily this was completed before the Covid-19 closures, which reduced the use of CBE's survey significantly.

After a survey is closed, CBE creates a report that includes a high-level summary of results and also detailed results for all questions. A "benchmarking scorecard" provides a comparison of the responses from the building's occupants to the broader CBE database, using categories that are averages of related questions. This approach is intended to anchor the individual building results to other buildings, as feedback among the various categories varies along predictable patterns. The scorecard for LWHS, shown in Figure 10.18, shows that the survey results are far superior to the benchmark averages, especially for the building overall, acoustics and cleanliness/maintenance.

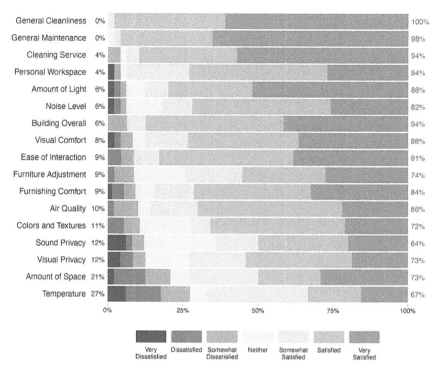

Figure 10.19 Occupant satisfaction is shown in the responses to key Likert-scale questions, ranked in terms of the level of satisfaction reported (percentages show the totals of three levels of dissatisfied and satisfied responses)

The report also provides detailed responses to all questions, for example, with bar charts similar to those shown in Figure 10.19. These positive survey results suggest that the design strategies used in this school renovation may be applicable to other older K-12 schools, which too often suffer from inadequate funding and poor maintenance.

Wider implications for survey-based building evaluations

These positive survey results also resulted in the project receiving the Livable Buildings Award in 2020. Conferred annually by CBE, this award was created to publicise the benefits of conducting POEs and to showcase buildings that provide high levels of occupant satisfaction as measured by the CBE survey, and that also excel in terms of sustainability and architectural design.[12]

Brad Jacobson, a partner and Chief Operating Officer for EHDD Architects, notes that feedback from the survey is valuable, and they strive to use it broadly across their work. However, they occasionally meet barriers to using it extensively. For example, as time is required between move-in and running a survey, design teams become focused on new projects, and client staff changes can occur. On the other hand, the POE process provides an opportunity to reconnect with clients and demonstrate the firm's commitment to creating projects that work for building users.

In other cases, design teams at EHDD have used survey results to identify and resolve problems. In the case of an advanced office project in Northern California, the survey revealed an issue related to glare; by using the responses to branching questions and text responses, the designers found that a gap at the edges of exterior motorised blinds was admitting direct sunlight directly to some work areas. They informed the client of this in a memo, and this led to a collaboration in which a minor retrofit was installed to resolve the issue. In another case, EHDD had designed a higher education building with an underfloor air distribution system, which allows building users to individually control the air supply at their workstations, intended to improve employees' thermal comfort. However, the survey found that comfort was not especially good, and people had never been informed that they could adjust the air diffusers on the floor. This allowed the building managers to correct this through better education of building users.

The survey results from LWHS, and the anecdotes about lessons learned on other EHDD projects, illustrate some of the benefits of web-based surveys. These can be executed cost-effectively by survey providers such as CBE and others and can also be part of a broader building evaluation, such as those prescribed by LEED, WELL, and the ASHRAE Performance Measurement Protocol, and may also include data collection from building management systems and physical measurements.

Notes

9 Additional information on the renovation is provided on the EHDD website at www.ehdd.com/project/lick-wilmerding-campus-expansion-and-renovation
10 Zagreus L., Huizenga C., Arens E. and Lehrer D. (2004) Listening to the occupants: A web-based indoor environmental quality survey. *Indoor Air*, 14(8), 65–74.
11 Graham L.T., Parkinson T. and Schiavon S. (2021) Lessons learned from 20 years of CBE's occupant surveys. *Buildings and Cities*, 2(1), 166–184.
12 Lehrer D. (2020) *Modernization of a Mid-Century High School Earns 2020's Livable Building Award*. Berkeley: CBE Centerline Blog, 7 December. https://cbe.berkeley.edu/centerline/modernization-of-a-mid-century-high-school-earns-2020s-livable-building-award/. (Sections of this case study are based on this post.)

11 Tried and tested POE methodologies

There are no hard and fast rules about which POE techniques should be used because each study will be unique to the organisation or prevailing situation. However, adopting well-known and widely used techniques does ensure that results are meaningful, that they can be repeated and therefore compared, and that the information can be benchmarked against other organisations that have used the same methodology. In my high-level review of established POE techniques, I found that:

- occupant feedback surveys and expert walkthroughs are widely used,
- questionnaires, particularly online ones, are more popular than interviews and focus groups; however, the latter is used more in the pre-project phase to inform the design,
- finding out how buildings perform compared with modelled predictions is an increasing objective of POEs,
- consequently, conducting energy and sustainability audits is more common,
- measurements of environmental conditions are less common in POE, but some (smart) buildings now incorporate environmental monitoring sensors,
- formal facilitated post-project discussions between project team members are less widely held, but there is an ongoing dialogue during the project,
- methods such as Soft Landings, which operate over the whole of the procurement cycle and on into aftercare, are rarely adopted in a commercial setting but they are adopted in UK public sector building projects.

This chapter summarises various generic methodologies that may be employed during POE (listed in alphabetical order) and that are still available at the time of publication. Nonetheless, there are many good legacy methods worthy of mention, but unfortunately no longer available, including: ABS Consulting's OLS; the BRE Checklist (*BRE Digest 478 Building Performance Feedback*); Housing Evaluation and Performance Studies (HEAPS); the OGC's Framework for Real-Estate Efficiency and Effectiveness (FREE); the Office Productivity Network's OPN Survey and OPN Index; and PROBE. Although the OPN no longer exists, I still use the occupant feedback survey that I developed for it.

Most of the established methods highlighted below emphasise occupant feedback, but some are more comprehensive. AMA WorkWare, for instance, includes an occupant survey, analysis of space efficiency, space-time utilisation studies

DOI: 10.1201/9781003350798-11

and user opinions captured through interviews and focus groups. Preiser and Vischer (2005) included examples of occupant feedback surveys in the appendices of *Assessing Building Performance*.

- **AMA WorkWare** – The POE toolkit of Alexi Marmot Associates (AMA) includes quantitative and qualitative methodologies which they invented and developed. AMA WorkWare provides a set of five techniques that may be used separately or in combination: user surveys, space-time utilisation studies, interviews, focus groups and space audits. AMA WorkWare can be used to create a brief and plan a project or as a post-project tool to review new or redesigned space. AMA WorkWare provides benchmarks of enduser opinion, space norms and utilisation patterns against its database of 60,000 people in over 250 office buildings. AMA WorkWareLEARN was later launched for educational buildings and has been used in over 100 buildings with 2,500 staff and student responses. www.aleximarmot.com/workware/

- **Building Occupants Survey System Australia (BOSSA)** – A suite of tools for assessing occupant satisfaction with IEQ and the performance of an office building. Tools include BOSSA Time-Lapse, a web-based survey tool; BOSSA Nova, a mobile IEQ assessment cart; and BOSSA Snap-Shot, a short questionnaire to assess the environment "right-here, right-now". BOSSA is endorsed for use in NABERS and by the Green Building Council of Australia and the Green Building Council of New Zealand. www.bossasystem.com/home.html

- **BSRIA Building Performance Evaluation** – BSRIA provides guidance on conducting BPEs and POEs and also offers a corresponding service. The organisation uses its BSRIA Occupant Wellbeing Survey to assess occupant satisfaction and wellbeing, covering the building's physical environment, indoor facilities, functionality and accessibility. In addition, it offers airtightness testing, U-value measurement, thermal imaging, energy metering strategy and analysis, forensic walkthrough, and IEQ monitoring. www.bsria.com/uk/consultancy/building-improvement/building-performance-evaluation/

- **BUS Methodology** – A questionnaire survey and benchmarking method created from thirty years of continuous development in building use studies. The occupant feedback survey was originally developed by Building Use Studies Ltd in 1985. The BUS Methodology helps to capture the complex nature of buildings without overwhelming occupants with too many questions. It highlights both good and less favourable building performance. The BUS Methodology is licensed through a network of selected partners who are trained to carry out high-quality surveys and interpret the results, to maintain the quality and integrity of the process. https://busmethodology.org.uk/index.html

- **CBE Occupant Survey** – The Center for the Built Environment (CBE) at the University of California, Berkeley, developed its cost-effective, web-based survey in 2000. The survey was originally developed as a research tool but is now used to collate employee feedback on its buildings. The survey has been implemented in over 1,000 buildings around the world, with responses from over 100,000 people. CBE also offers other complementary POE services, such as IEQ monitoring. https://cbe.berkeley.edu/resources/occupant-survey/

- **Clinic Design Post-Occupancy Evaluation Toolkit** – The Center for Health Design in California has developed paper-based surveys to collate feedback from staff and patients in healthcare environments. www.healthdesign.org/insights-solutions/clinic-design-post-occupancy-evaluation-toolkit-pdf-version

- **Comfortmeter** – A web-based survey tool for assessing user satisfaction in office buildings that enables facility managers to diagnose problems in a building. There are two standard versions of Comfortmeter, but a tailored questionnaire can be developed based on a customer's needs. Comfortmeter – Indoor Environmental Quality focuses on the indoor environment of a workspace, including thermal comfort, acoustics and quality of air and light. Comfortmeter – User Well-being is a holistic survey that goes beyond IEQ and takes into account other factors that impact the health and wellbeing of building users. www.comfortmeter.eu/

- **Cruxera** – An innovative solution that emphasises co-creation and employee empowerment. By embracing digital transformation and harnessing the power of business intelligence, it enables organisations to progress from descriptive to predictive analytics. With its comprehensive data management and analytics platform, Cruxera empowers organisations to gain valuable insights into employee usage, optimise occupancy rates, and make informed decisions regarding resource allocation. The Cruxera platform includes questionnaires, sensors (air quality, light, occupancy, density), booking systems reports, access control reports, and a consulting option for management interviews and workshops. http://cruxera.io/

- **Design Quality Index (DQI)** – In 1999, the Commission for Architecture and the Built Environment (CABE), the Department of Trade and Industry (DTI), the Office of Government Commerce (OGC), Constructing Excellence and the Strategic Forum of Construction came together to develop the DQI as a means of addressing the issue of poor-quality design in buildings. DQI is a structured assessment procedure for assessing the three core principles of design: functionality (*utilitas*), build quality (*firmitas*) and impact (*venustas*). It is a facilitated process that takes the form of workshops supported by a bespoke evaluation model and an independent, accredited facilitator. www.dqi.org.uk/

- **Design Quality Method (DQM)** – BRE, based in the UK, has a long history of offering POE services as well as environmental assessment. Its DQM assesses design quality, building performance feedback and operational efficiency using expert opinion, user views and scientific measurement. DQM covers all aspects of building performance, including architecture, environmental engineering, user comfort, whole-life costs, detailed design and user satisfaction. The DQM is mostly used by public sector bodies, and the results are benchmarked against the 100-building database and industry standards.
www.bre.co.uk/page.jsp?id=1623

- **Healthy Building Index (HBI)** – Developed by the indoor environmental consultancy BBA, the HBI tool is used for determining employee satisfaction with the indoor environment and with the facilities of the work environment. The calculated level of satisfaction is expressed using a number between 100 and 1,000, which represents the building HBI. The building result is then compared against the performance of an average office building. The HBI is based on six themes: thermal comfort, indoor air quality, light, acoustics and facilities (ergonomics, health promotions and cleanliness).
https://healthybuildingindex.nl/

- **Leesman Index (Lmi)** – Launched in 2020, Leesman uses its standardised online surveys to measure and benchmark employee workplace experience. The surveys have standard modules and variations, including surveys for the office, home, hybrid and inside. The Lmi represents an overall score of the building performance based on occupant experience. Leesman has conducted surveys in over 6,500 buildings across 100+ countries with over 1,000,000 responses. This makes it the largest user experience database, allowing global benchmarking across business sectors.
www.leesmanindex.com/.

- **National Environmental Assessment Toolkit (NEAT)** – The Center for Building Performance and Diagnostics at Carnegie Mellon University developed a unique and proprietary method for evaluating the impact of environmental conditions on occupants. The method consists of five tools: environmental instrumentation, occupant satisfaction/collaboration survey, technical attributes of building systems survey, real-time physical indicator identification software and central database systems.
https://soa.cmu.edu/bpd

- **Occupant Comfort & Wellness Survey** – The Institute for the Built Environment (IBE) at Colorado State University offers basic and customisable occupant survey services, as well as focus groups, interviews, social network analysis and programming support services. IBE's current survey was developed for office buildings, and IBE is developing a multi-family residential version. Survey

topics include acoustics, air quality, lighting, thermal comfort, layout, nutrition, nature, sleep, productivity and wellness.
https://ibe.colostate.edu/occupant-comfort-wellness-surveys/

- **PeopleLOOK** – Baker Stuart's web-based staff engagement/staff satisfaction survey is designed to assess the perceived satisfaction of employees along with areas for improvement. It also incorporates psychometric elements to review in-depth the personality profile of different teams to enable future workplaces to better match team needs. As well as understanding employees' views on the importance and satisfaction of various elements of their workspace, working lives and wellbeing, PeopleLOOK also explores the breakdown of activities and time in various locations such as the office, home and other remote locations. It is a comprehensive survey with additional modules covering team personality, wellbeing, location, travel and organisational culture. Baker Stuart also offers SpaceLOOK, a space-utilisation survey, and FileLOOK, a filing and storage audit.
https://bakerstuart.com/peoplelook/

- **School Building Assessment Methods** – Developed by Henry Sanoff in 2001, this manual of tools includes feedback surveys, photo questionnaires, walk-through checklists, observation forms and small-group discussion tools. The methods are designed to encourage stakeholders to discover and reflect on the physical features of school buildings. The methods cover spatial arrangements, the indoor environment, architecture, wayfinding, cleanliness and maintenance and so on.
www.academia.edu/12023027/School_Building_Assessment_Methods

- **SmartSite KPIs** – This is an online tool developed by Constructing Excellence and BRE. It differs from the other methods in this chapter because it focuses on the construction phase rather than post-project. The tool offers benchmarking of the performance of construction projects against the rest of the construction industry using the established and nationally recognised 120+ KPIs.
www.bresmartsite.com/products/smartsite-kpis/

- **Space Performance Evaluation Questionnaire (SPEQ)** – Developed by High Performance Environments Lab (HiPE), University of Oregon, this questionnaire has been tested in a variety of building types since 1998 and has been applied to evaluate more than 150 buildings with a robust database containing more than 100,000 data points. This allows responses to be benchmarked against a comparative baseline. The online occupant survey has developed categories and scales representative of the most important issues identified by occupants as impacting their comfort, satisfaction, performance, health and wellbeing. HiPE also conducts environmental monitoring.
https://blogs.uoregon.edu/hipetest/about/

- **Sustainable and Healthy Environments (SHE)** – Developed by the University of Melbourne, SHE is a powerful data-collection vehicle feeding ongoing research projects about the performance of spaces from the occupants' perspective. The SHE survey collects data about human, organisational and environmental-related variables that, combined, may affect occupants' satisfaction, health (physical and emotional) and productivity. Surveys can be used for performance evaluation of workplaces, schools and precincts.
  https://msd.unimelb.edu.au/she/research/she-post-occupancy-evaluation-survey

- **Tenant Satisfaction Measures** – The UK Regulator of Social Housing is currently developing a new system for assessing social housing landlords on their provision of good quality homes and services. Property owners letting 1,000 or more homes will be required to submit their tenant satisfaction measures data in 2024. The key measurement tool is an occupant survey, including questions on satisfaction overall and with maintenance, repairs and occupant engagement.
  www.gov.uk/government/consultations/consultation-on-the-introduction-of-tenant-satisfaction-measures/outcome/annex-5-tenant-satisfaction-measures-tenant-survey-requirements-accessible

- **WPU Occupant Feedback Survey** – Nigel Oseland of Work*place* Unlimited (WPU) has conducted over 100 POEs of offices and education facilities. The POE methodology primarily consists of an occupant feedback survey, originally developed for the OPN and made public by Oseland and Bartlett (1999). The questionnaire is tailored for different environments and circumstances, but the standardised questions and database of over 10,000 responses allow some benchmarking. The questionnaire is usually accompanied by an expert walk-through, and further feedback is obtained through interviews and focus groups. Other metrics, such as space analysis and utilisation surveys, are often included in the POE.
  https://workplaceunlimited.com/services/evaluation-feedback.html

- **ZZA UX:SPACE** – ZZA Responsive User Environments is a research and strategy practice supporting the development of people-centric, sustainable buildings and places. Working across typologies, ZZA works at the nexus of social, physical and virtual domains, linking cultural, design and management perspectives to help realize the positive potentials of any project and urban space. Founded and directed by Ziona Strelitz, ZZA uses distinctive methodologies of design anthropology in rigorous research on building use and the implications of continuously evolving technology and culture for the social functioning and design of space. Its distinctive focus on user experience is connoted in its marks, UX:SPACE®, UX:WORKPLACE® and UX:CAMPUS®.
  www.zza.co.uk/what/

Epilogue

This book captures my thoughts and experience in conducting POEs. It is intended as a practical guide and useful resource for those new to POE and a timely reminder of its relevance to those already familiar with the process. I have included guidance on various POE techniques, some fundamental to POE and others more useful in detailed studies of the impact of buildings on occupants. I have attempted to demystify POE, clarify the benefits of conducting one and disband the common barriers. My primary intention is that this book will promote more uptake of POE in various relevant formats.

Understanding the value and performance of any building project is a worthy pursuit but more so for offices. Post-pandemic, the purpose and benefits of offices are being questioned more than ever, and it is increasingly important to demonstrate that offices support the occupying organisation and individuals, enhancing their performance and wellbeing. The impact that building and operating an office has on the environment is also more relevant than ever, so it is crucial that an office performs as designed and modelled. Also, in times of austerity it is more crucial than ever to justify and prove the value of a building project. Evaluating a new or refurbished office after it is completed, and the occupants have moved in, is key to understanding these queries.

POE can be complex and off-putting, but a basic POE consisting of a few core elements is feasible and cost-effective. Undertaking a POE need not be an overwhelming task – it is better that a simple POE is carried out rather than completely ignoring the impact of the building on its occupants and the environment. Revisit and evaluate a recent building project and the benefits of conducting a POE – for the occupants, the organisation and the project team – will become clear.

I believe it is the responsibility of all those in the real estate, design and construction industries to conduct POEs on behalf of their clients. POEs verify that quality and value have been delivered and help ensure the quality and value of future buildings.

I hope this book will inspire you to take your first, or further, steps in post-occupancy evaluation.

Bibliography

Abdou A. and Al Dghaimat M. (2016) *Post Occupancy Evaluation of Educational Buildings: A Case Study of a Private School in the UAE*. Proceedings of 4th Annual International Conference on Architecture and Civil Engineering, https://www.scribd.com/document/541948820/Post-Occupancy-Evaluation-of-Educational-Buildings.

Agha-Hossein M., Birchall S. and Vatal S. (2015) *Building Performance Evaluation in Non-Domestic Buildings*. Bracknell: Building Services Research and Information Association.

Allen T., Bell A., Graham R., Hardy B. and Swaffer F. (2004) *Working Without Walls: An Insight into the Transforming Government Workplace*. London: Office of Government Commerce.

ASHRAE (2013) *Handbook for Workplace Acoustic Quality Assessment*. Atlanta: The American Society of Heating, Refrigerating and Air-Conditioning Engineers.

ASHRAE (2020) *ANSI/ASHRAE Standard 55–2020, Thermal Environmental Conditions for Human Occupancy*. Atlanta: The American Society of Heating, Refrigerating and Air-Conditioning Engineers.

ASHRAE (2022) *ANSI/ASHRAE Standard 62.1–2022, Ventilation and Acceptable Indoor Air Quality*. Atlanta: The American Society of Heating, Refrigerating and Air-Conditioning Engineers.

BCO (2011) *Guide to Fit Out*. London: British Council for Offices.

BCO (2014) *Guide to Specification*. London: British Council for Offices.

BIFM (2016) *Operational Readiness Guide*. Bishop's Stortford: British Institute of Facilities Management.

Bordass B. and Leaman A. (2005) Making feedback and post-occupancy evaluation routine 1: A portfolio of feedback techniques. *Building Research & Information*, 33(4), 361–375.

BRE (2016) *BREEAM In-Use International Technical Manual*. Watford: BRE Global Limited.

BSI (2014a) *BS 8233 Guidance on Sound Insulation and Noise Reduction for Buildings*. London: British Standards Institution.

BSI (2014b) *BS EN 13779 Ventilation for Non-residential Buildings – Performance Requirements for Ventilation and Room-Conditioning Systems*. London: British Standards Institution.

BSI (2014c) *PAS 1192–3 Specification for Information Management for the Operational Phase of Assets Using Building Information Modelling*. London: British Standards Institution.

BSI (2022a) *BS 40101 Building Performance Evaluation of Occupied and Operational Buildings*. London: British Standards Institution.

BSI (2022b) *BS 8536 Design, Manufacture and Construction for Operability*. London: British Standards Institution.

BSRIA (2009) *The Soft Landings Framework for Better Briefing, Design, Handover and Building Performance In-Use*. Bracknell: Building Services Research and Information Association.

Bunn R. (2020) *TM62 Operational performance: Surveying Occupant Satisfaction*. London: Chartered Institution of Building Services Engineers.

CABE (2008) *Sure Start Children's Centres: A Post-Occupancy Evaluation*. London: Commission for Architecture and the Built Environment.

CABE and BCO (2006) *The Impact of Office Design on Business Performance*. London: The Commission for Architecture and the Built Environment and British Council for Offices.

Carthey J. (2006) Post occupancy evaluation: Development of a standardised methodology for Australian health projects. *International Journal of Construction Management*, January, 57–74.

CEN (2021) *EN 12464–1 Light and Lighting – Lighting of Work Places – Part 1: Indoor Work Places*. Brussels: Comité Européen de Normalisation.

Cherry K. (2022) What is the negativity bias? *Verywell Mind*, November.

CIBSE (2006) *TM22 Energy Assessment and Reporting Methodology*. London: Chartered Institution of Building Services Engineers.

CIBSE (2008) *TM46 Energy Benchmarks*. London: Chartered Institution of Building Services Engineers.

CIBSE (2015a) *Guide A4 Environmental Design*. London: Chartered Institution of Building Services Engineers.

CIBSE (2015b) *TM57 Integrated School Design*. London: Chartered Institution of Building Services Engineers.

CIBSE (2022) *TM23 Testing Buildings for Air Leakage*. London: Chartered Institution of Building Services Engineers.

Cohen R., Standeven M., Bordass W. and Leaman A. (2001) Assessing building performance in use 1: The PROBE process. *Building Research and Information*, 29(2), 85–102.

Concerto Consulting (2006) *Getting the Best from Public Sector Office Accommodation: Case Studies*. London: National Audit Office.

Connell B.R. and Ostrander E.R. (1976) *Methodological Considerations in Post Occupancy Evaluation: An Appraisal of the State of the Art*. Washington, DC: The American Institute of Architects Research Corporation.

Constructing Excellence (2018) *UK Industry Performance Report 2018*. London: Glenigan.

Cooper I. (2001) Post-occupancy evaluation – where are you? *Building Research and Information*, 29(2), 158–163.

Department of Housing and Public Works (2017) *Strategic Asset Management Framework: Post Occupancy Evaluation*. The State of Queensland: Department of Housing and Public Works.

Department of Housing and Urban Development (1977) *Post Occupancy Evaluations of Residential Environments: An International Bibliography*. Washington, DC: Department of Housing and Urban Development, Office of Policy Development and Research.

DETR (2000) *KPI Report for the Minister for Construction by the KPI Working Group*. London: Department of the Environment, Transport and the Regions.

DETR (2003) *Energy Consumption Guide 19: Energy Use in Offices*. London: Department of the Environment, Transport and the Regions.

Deuble M.P. and de Dear R.J. (2014) Is it hot in here or is it just me? Validating the post-occupancy evaluation. *Intelligent Buildings International*, 6(2), 112–134.

Egan J. (1998) *Rethinking Construction*. London: Department of Trade and Industry.

Enright S. (2002) Post-occupancy evaluation of UK library building projects: Some examples of current activity. *Liber Quarterly*, 12, 26–45.

Fairley P. (2015) Feedback loop: With a frequent gap between predicted and actual performance, post-occupancy evaluations begin to catch on. *Architectural Record*, 16 August.

Fitwel (2021) *Fitwel Enhanced Indoor Air Quality Testing Policy*. New York: The Center for Active Design.

Fletcher P. and Satchwell H. (2015) *Briefing: A Practical Guide to RIBA Plan of Work 2013 Stages 7, 0 and 1*. London: Royal Institute of British Architects.

Fowler K.M., Rauch E.M., Henderson J.W. and Kora A.R. (2011) *Re-Assessing Green Building Performance: A Post Occupancy Evaluation of 22 GSA Buildings*. Richland: Pacific Northwest National Laboratory.

Friedman A., Zimring C. and Zube C. (1978) *Environmental Design Evaluation*. New York: Plenum.

Gillen N. (2015) Productivity: How workplace design can make a difference. In *Infrastructure Intelligence*. London: Association for Consultancy and Engineering, August.

Guilford J.P. (1947) The discovery of aptitude and achievement variables. *Science*, 6(2752), 279–282.

Gupta R. and Gregg M. (2020) *State of the Nation Review: Performance Evaluation of New Homes*. Oxford: Building Performance Network and Oxford Brookes University.

Hadi M. and Kiruthiga S.V. (2008) *Post Occupancy Evaluation of the Sheffield International College, University of Sheffield, Client Report Number, 123–615*. Watford: Building Research Establishment.

Hadjri K. and Crozier C. (2009) Post-occupancy evaluation: Purpose, benefits and barriers. *Facilities*, 27(1–2), 21–33.

Hassanain M.A. and Mudhei A.A. (2006) Post-occupancy evaluation of academic and research library facilities. *Structural Survey*, 24(3), 230–239.

Heath O., Jackson V., Goode E. and Hadi M. (2019) *Creating Positive Spaces – Measuring the Impact of Your Design*. Atlanta: Interface.

Heerwagen J. and Zagreus L. (2005) *The Human Factors of Sustainable Building Design: Post Occupancy Evaluation of the Philip Merrill Environmental Center*. Berkeley: Center for the Built Environment.

HEFCE/AUDE (2006) *Guide to Post Occupancy Evaluation*. London: Higher Education Funding Council for England.

Hillier B. and Hanson J. (1984) *The Social Logic of Space*. Cambridge: Cambridge University Press.

Home Office (2009) *Design Policy – North Kent Police Station, Post Occupancy Evaluation*. London: Home Office.

IFMA (2012) *Workplace Amenities Strategies, Research Report #36*. Houston: International Facility Management Association.

Ikediashi D., Udo G. and Ofoegbu M. (2020) Post-occupancy evaluation of University of Uyo buildings. *Journal of Engineering, Design and Technology*, 18(6), 1711–1730.

Institute of Medicine (2015) *Psychological Testing in the Service of Disability Determination*. Washington, DC: The National Academies Press.

IPA (2021) *Gate 5 Review: Operations Review and Benefits Realisation*. London: Infrastructure and Projects Authority.

IPD (2008) *IPD Environment Code: Measuring the Environmental Performance of Buildings*. London: Investment Property Databank.

IPD (2013) *Global Estate Measurement Code for Occupiers*. London: Investment Property Databank.

ISO (2005) *ISO 7730:2005 Ergonomics of the Thermal Environment – Analytical Determination and Interpretation of Thermal Comfort Using Calculation of the PMV and PPD Indices and Local Thermal Comfort Criteria.* Geneva: International Organization for Standardization.

ISO (2018) *ISO 19650 Organization and Digitization of Information About Buildings and Civil Engineering Works, Including Building Information Modelling (BIM).* Geneva: International Organization for Standardization.

ISO (2021) *ISO 22955 Acoustics – Acoustic Quality of Open Office Spaces.* Geneva: International Organization for Standardization.

IWBI (2016) *The WELL Building Standard v1.* New York: International WELL Building Institute.

IWBI (2022) *The WELL Building Standard v2.* New York: International WELL Building Institute.

IWFM (2017) *Employer's Information Requirements (EIR).* Bishop's Stortford: Institute of Workplace and Facilities Management.

John O.P. and Srivastava S. (1999) The big five trait taxonomy: History, measurement, and theoretical perspectives. In L.A. Pervin and O.P. John (eds) *Handbook of Personality: Theory and Research.* New York: Guilford Press, 102–138.

Kalantari S. and Snell R. (2017) Post-occupancy evaluation of a mental healthcare facility based on staff perceptions of design innovations. *HERD*, 10(4), 121–135.

Kennett E. (2021) Post occupancy evaluations. In *WBDG – Whole Building Design Guide.* Washington, DC: National Institute of Building Sciences.

Lackney J.A. and Zajfen P. (2005) Post-occupancy evaluation of public libraries: Lessons learned from three case studies. *Library Administration & Management*, 19(1), 16–25.

Leesman (2015) *Leesman Review*, Issue 19. www.leesmanindex.com/leesman-review/.

Loftness V., Aziz A., Choi J.H., Kampschroer K., Powell K., Atkinson M. and Heerwagen J. (2009) The value of post-occupancy evaluation for building occupants and facility managers. *Intelligent Buildings International*, 1, 1–20.

Loftness V., Hartkopf V., Aziz A., Choi J.-H. and Park J. (2018) Critical frameworks for building evaluation: User satisfaction, environmental measurements and the technical attributes of building systems (POE + M). In W.F.E. Preiser, A.E. Hardy and U. Schramm (eds) *Building Performance Evaluation.* Cham: Springer International, 29–48.

Markus T.A. (1967) The role of building performance measurement and appraisal in design method. *Architects' Journal*, 146(25), 1565–1573.

Markus T.A. (1972) *Building Performance.* London: Applied Science Publishers.

Mashford K. and Gill Z. (2022) *Launch of the British Standard for In-use Building Performance Evaluation.* Presentation. https://building-performance.network/advocacy/british-standard-bs40101-launch.

McLaughlin H. (1975) Post-occupancy evaluation of hospitals. *American Institute of Architects Journal*, January, 30–34.

Miller G.A. (1956) The magical number seven, plus or minus two: Some limits on our capacity for processing information. *Psychological Review*, 63(2), 81–97.

Murphy M. (2015) Which of these 4 communication styles are you? *Forbes*, August.

NSW Treasury (2004) *10 Steps to Procurement Process – Construction.* New South Wales: New South Wales Treasury.

OGC (2007) *Improving Performance Project Evaluation and Benchmarking: Achieving Excellence in Construction Procurement Guide.* London: Office of Government Commerce.

Oseland N.A. (1995) Predicted and reported thermal sensation in climate chambers, offices and homes. *Energy and Buildings*, 23, 105–115.

Oseland N.A. (2007) *Guide to Post-Occupancy Evaluation*. London: BCO.

Oseland N.A. (2018) From POE to BPE: The next era. In W.F.E. Preiser, A. Hardy and U. Schramm (eds) *Building Performance Evaluation: From Delivery Process to Life Cycle Phases*. Cham: Springer International.

Oseland N.A. (2022) *Beyond the Workplace Zoo: Humanising the Office*. London: Routledge.

Oseland N.A. and Bartlett P. (1999) *Improving Office Productivity: A Guide for Business and Facilities Managers*. Harlow: Pearson Education Ltd.

Oseland N.A., Bunn R., Oldman T. and Hadcocks M. (2022) *The Future of UK Office Densities*. London: British Council for Offices.

Oseland N.A. and Hodsman P. (2017) Psychoacoustics: Resolving noise distractions in the workplace. Chapter 4 in A. Hedge (ed) *Ergonomics Design for Healthy and Productive Workplaces*. Abingdon: Taylor & Francis.

Oseland N.A., Tucker M. and Wilson H. (2023) Developing the return on workplace investment (ROWI) tool. *Corporate Real Estate Journal*, 12(2), 1–13.

Ozturk Z., Arayici Y. and Coates, S. (2012) *Post Occupancy Evaluation (POE) in Residential Buildings Utilizing BIM and Sensing Devices: Salford Energy House Example*. Presented at Retrofit, The Lowry, Salford Quays, January.

Palmer J., Terry N. and Armitage P. (2016) *Building Performance Evaluation Programme: Findings from Non-domestic Projects – Getting the Best from Buildings*. London: Innovate.

Park J., Loftness V. and Aziz A. (2018) Post-occupancy evaluation and IEQ measurements from 64 office buildings: Critical factors and thresholds for user satisfaction on thermal quality. *Buildings*, 8(11), 156.

Preiser W.F.E. (1994) Built environment evaluation: Conceptual basis, benefits and uses. *Journal of Architectural and Planning Research*, 11(2), 91–107.

Preiser W.F.E., Davis A.T., Salama A.M. and Hardy A. (2015) *Architecture Beyond Criticism: Expert Judgment and Performance Evaluation*. New York: Routledge.

Preiser W.F.E., Hardy A. and Schramm U. (2018) *Building Performance Evaluation: From Delivery Process to Life Cycle Phases*. Cham: Springer International.

Preiser W.F.E., Rabinowitz H. and White E. (1988) *Post-Occupancy Evaluation*. New York: Van Nostrand Reinhold.

Preiser W.F.E. and Schramm U. (1997) Building performance evaluation. In D. Watson et al. (eds) *Time-Saver Standards: Architectural Design Data*. New York: McGraw-Hill.

Preiser W.F.E. and Schramm U. (2005) A conceptual framework for building performance evaluation. In W.F.E. Preiser and J.C. Vischer (eds) *Assessing Building Performance*. Oxford: Elsevier Butterworth-Heinemann.

Preiser W.F.E. and Vischer J.C. (2005) *Assessing Building Performance*. Oxford: Elsevier Butterworth-Heinemann.

Randall, T. (2015) The smartest building in the world: Inside the connected future of architecture. *Bloomberg Businessweek*, September 23.

RIBA (1964) *RIBA Plan of Work*. London: Royal Institute of British Architects.

RIBA (1991) A research report for the architectural profession. In F. Duffy and L. Hutton (eds) *Architectural Knowledge: The Idea of a Profession*. London: Taylor & Francis.

RIBA (2013) *RIBA Plan of Work 2013*. London: Royal Institute of British Architects.

RIBA (2016) *Post Occupancy Evaluation and Building Performance Evaluation Primer*. London: Royal Institute of British Architects.

RIBA (2017) *Building Knowledge: Pathways to Post Occupancy Evaluation.* London: Royal Institute of British Architects.

RIBA (2020a) *RIBA Plan of Work 2020.* London: Royal Institute of British Architects.

RIBA (2020b) *Post Occupancy Evaluation: An Essential Tool to Improve the Built Environment.* London: Royal Institute of British Architects.

RICS (2013) *Cost Analysis and Benchmarking, 1st Edition, Guidance Note.* London: Royal Institution of Chartered Surveyors.

Sailer K. (2010) *The Space-organisation Relationship: On the Shape of the Relationship Between Spatial Configuration and Collective Organisational Behaviours.* Technical University of Dresden.

Sailer K., Budgen A., Lonsdale N., Turner A. and Penn A. (2010) Pre and post occupancy evaluations in workplace environments: Theoretical reflections and practical implications. *The Journal of Space Syntax,* 1(1), 199–213.

Salami A.R., Akande I. and Oke T.I. (2022) Post-occupancy evaluation (POE) of shopping malls in Ota, Ogun State, Nigeria. *International Journal of Innovative Science and Research Technology,* 7, 1773–1776.

Saldaña J. (2015) *The Coding Manual for Qualitative Researchers.* Thousand Oaks: Sage Publishing.

Sanni-Anibire M.O., Hassanain M.A. and Al-Hammad A.-M. (2016) Post-occupancy evaluation of housing facilities: Overview and summary of methods. *Journal of Performance of Constructed Facilities,* 30(5).

Schiller G., Arens E., Bauman F., and Benton C. (1988) A field study of thermal environments and comfort in office buildings. *ASHRAE Transactions,* 94(2), 208–308.

Schoenefeldt H. (2019) The house of commons: A precedent for post-occupancy evaluation. *Building Research & Information,* 479(6), 635–665.

Skills Funding Agency (2014) *Post-Occupancy Evaluation Guide.* London: Skills Funding Agency.

Stevenson F. (2019) Six steps towards effective post occupancy evaluation for homes. *The RIBA Journal,* October.

Stroop J.R. (1935) Studies of interference in serial verbal reactions. *Journal of Experimental Psychology,* 18(6), 643–662.

Thomazoni A.D.A.L., Ornstein S.W. and Ono R. (2016) Post-occupancy evaluation applied to the design of a complex hospital by means of the flow analysis. *Proceedings of 50th International Conference of the Architectural Science Association,* 537–546.

Turner C. (2006) *LEED Building Performance in the Cascadia Region: A Post Occupancy Evaluation Report.* Seattle: Cascadia Region Green Building Council.

UK BIM Framework (2019) *Government Soft Landings: Revised Guidance for the Public Sector on Applying BS8536 Parts 1 and 2, Updated for ISO 19650.* London: UK BM Framework.

US Green Building Council (2019) *LEED v4 for Interior Design and Construction.* Washington, DC: US Green Building Council.

Usher N. (2018) *The Elemental Workplace: The 12 Elements for Creating a Fantastic Workplace for Everyone.* London: LID Publishing Ltd.

Van der Ryn S. and Silverstein M. (1967) *Dorms at Berkeley: An Environmental Analysis.* Berkeley: Center for Planning and Development Research, University of California.

Watson C. and Thomson K. (2005) Bringing post-occupancy evaluation to schools in Scotland. *Evaluating Quality in Educational Facilities,* 3, 189–220.

Watson P. (2020) *News and Thoughts: The Uncomfortable Truth About Post-Occupancy Evaluation.* London: HLM Architects.

Way M. and Bordass B. (2005) Making feedback and post-occupancy evaluation routine 2: Soft landings – involving design and building teams in improving performance. *Building Research & Information*, 33(4), 353–360.

Wener R.E. (1994) *Post Occupancy Evaluation Procedure: Instruments and Instructions for Use*. Washington, DC: The National Institute of Corrections.

Whitemyer D. (2006) Anthropology in design: An old science makes an impact on interior design. *IIDA Perspective*. Spring, London.

Wilson D. (2006) *BOP:* Making sense of space. Proceedings of *IET Wireless Sensor Networks Conference*, London, December.

Zimmerman A. and Martin M. (2001) Post-occupancy evaluation: Benefits and barriers. *Building Research and Information*, 29, 168–174.

Zuo J., Yuan X. and Pullen S.F. (2011) Post occupancy evaluation study in hospital buildings – a pilot study. *Applied Mechanics and Materials*, 94–96, 2248–2256.

Index

Note: Page numbers in *italic* indicate a figure on the corresponding page.

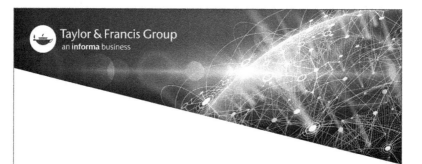

Milton Keynes UK
Ingram Content Group UK Ltd.
UKHW020206091124
450853UK00015B/180